Adaptations of Desert Organisms

Edited by J. L. Cloudsley-Thompson

Springer-Verlag Berlin Heidelberg GmbH

Volumes published in the series

Ecophysiology of the Camelidae and Desert Ruminants
By R. T. Wilson (1989)

Ecophysiology of Desert Arthropods and Reptiles
By J. L. Cloudsley-Thompson (1991)

Plant Nutrients in Desert Environments
By A. Day and K. Ludeke (1993)

Seed Germination in Desert Plants
By Y. Gutterman (1993)

Behavioural Adaptations of Desert Animals
By G. Costa (1995)

Invertebrates in Hot and Cold Arid Environments
By L. Sømme (1995)

Energetics of Desert Invertebrates
By H. Heatwole (1996)

Ecophysiology of Desert Birds
By G. L. Maclean (1996)

Plants of Desert Dunes
By A. Danin (1996)

Biotic Interactions in Arid Lands
By J. L. Cloudsley-Thompson (1996)

Structur-Function Relations of Warm Desert Plants
By A. C. Gibson (1996)

Physiological Ecology of North American Desert Plants
By S. D. Smith, R. K. Manson, and J. E. Anderson (1997)

Ecophysiology of Small Desert Mammals
By A. A. Degen (1997)

Homeostasis in Desert Reptiles
By S. D. Bradshaw (1997)

Ecophysiology of Amphibians Inhabiting Xeric Environments
By M. R. Warburg (1997)

Avian Desert Predators
By W. E. Cook (1997)

Carnivore Ecology in Arid Lands
By J. du P. Bothma (1998)

Ecophysiology of Economic Plants in Arid and Semi-Arid Lands
By G. E. Wickens (1998)

Arthropods of Mediterranean-Type Ecosystems
By G. P. Stamou (1998)

Dispersal Biology of Desert Plants
By K. van Rheede van Oudtshoorn and M. W. van Rooyen (1999)

Spatial Ecology of Desert Rodent Communities
By G. I. Shenbrot, B. R. Krasnov, and K. A. Rogovin (1999)

Desert Arthropods: Life History Variations
By F. Punzo (2000)

Plants in the Deserts of the Middle East
By K. H. Batanouny (2001)

Kamal H. Batanouny

Plants in the Deserts of the Middle East

With 74 Figures and 8 Tables

Springer

Prof. Dr. Kamal H. Batanouny
University of Cairo
Faculty of Science
Giza
Egypt

Cover illustrations: Acacia raddiana, South Sinai (front cover) and *Asphodelus ramosus*, Mediterranean coastal zone, Egypt (front and back cover). Photographs taken by Kamal H. Batanouny

ISSN 1430-9432
ISBN 978-3-642-08092-0

Library of Congress Cataloging-in-Publication Data.

Batanouny, K. H., 1936-
Plants in the deserts of the Middle East / Kamal H. Batanouny.
p. cm. - (Adaptations of desert organisms)
Includes bibliographical references (p.).
ISBN 978-3-642-08092-0 ISBN 978-3-662-04480-3 (eBook)
DOI 10.1007/978-3-662-04480-3
1. Desert plants–Middle East. 2. Desert plants–Adaptation–Middle East. I. Title. II. Series
QK379.5.M53 B37 2000
581.7"54'0956–dc21 00-061191

Originally published by Springer-Verlag Berlin Heidelberg **New York in** 2001
Softcover reprint of the hardcover 1st edition 2001

Cover design: *design & production* GmbH, Heidelberg
Typesetting: Bader • Damm • Kröner, Heidelberg

SPIN 10022240 31/3130 YK - 5 4 3 2 1 0 - Printed on acid free paper

بسم الله الرحمن الرحيم

" وجعلنا من الماء كل شيء حي "

من الآية ٣٠-الأنبياء

القرآن الكريم

In the Name of God, Most Gracious, Most Merciful
"We Made from Water Every Living Thing"

From Verse 30, Sura XXI, Anbya'a
The Holy Qura'an

Acknowledgements

This book could not have been written without the direct and indirect help of many people. Above all, I owe a debt of gratitude to my late mother who contributed a great deal to my life. The continuous support and advice, until now, of my father is also highly appreciated. In addition, I would like to express my sincere gratitude to all those who have in one way or another taught, coached, guided, encouraged, and most important, corrected me in some 56 years of education and learning. Most sincere appreciation to the Bedouin in the deserts of the Middle East, from whom I have learned, during my several trips to these deserts, more than anyone could have expected.

My professors who introduced me to plant ecology and ecophysiology influenced me to a great extent, an influence for which I cannot be grateful enough. I cannot mention them all by name, but it is impossible to forget the profound influence of Professor A.A. Abdel Rahman, Professor H. Ziegler and of the late Professor A.M. Migahid.

I wish to thank Professor J.L. Cloudsley-Thompson for his invaluable comments and careful editing of this book. I am sure that this book will remind him of his trips with me to the desert of Qatar, where we enjoyed an exchange of knowledge about and many thoughts on the plant life existing there under the harsh desert conditions.

Finally, I wish to express my heartfelt thanks to my wife for her unrelenting support and for her patience, understanding, tolerance and valuable assistance.

Giza, Autumn 2000 K.H. Batanouny

Contents

CHAPTER 1

Introduction

Usually authors write introductions for their books, although they know that not many readers will read it. Despite this, authors insist on writing an introduction and no publisher will publish a book without one.

I would like to inform my dear readers that I have spent almost all of the first quarter of my life in a village in the Nile Delta, 65 km north of Cairo. The everyday scenery there was the beautiful green landscape dissected with canals full of running water. All of these were bordered with the huge sycamore, mulberry and acacia trees. The desert was something unknown to me at that time, except for the very basic information given in geography books, which explained that the desert is a place without water or cultivation. Some of my ideas about the desert came to me from the stories in the history of Islam and the desert lands where Islam originated.

My real attraction to the desert developed in the last year of my undergraduate studies. This was during the field courses in Ecology (Prof. A.M. Migahid and Prof. A.A. Abdel Rahman) and Desert Flora (Prof. V. Taeckholm). With this background, I came to believe that the vast desert, which occupies about 96 % of my country, is the best field of study for a biologist. I also firmly believed that Egyptians should contribute to scientific studies of the deserts, which represent the main gate to their future in view of the ever-increasing population in the narrow Delta and Nile Valley.

From 1956 until now, I have been engaged in studies of the deserts of Egypt and of other countries in the Middle East. When I received the Alexander von Humboldt fellowship, I made use of the facilities in the laboratories of the Technical University in Munich to investigate the ecophysiology of desert plants. This reminds me of an interesting story: When, in 1971, the Alexander von Humboldt fellows were invited to visit the President of the German Federal Republic in Bonn (Dr. Dr. Heinemann). I introduced myself as an Egyptian botanist and he was pleased to tell me that his wife loved flowers, especially roses. He was astonished when I told him that I study desert plants, and exclaimed: but we have no deserts in Germany! My answer was: "*Ich habe die Samen mitgebracht.*"

All my life, I have roamed the deserts of the world with the desert ecology in my mind and the desert plants in my heart. So, when I was invited by Springer-Verlag and Prof. J. Cloudsley-Thompson to contribute to the series on the adaptation of desert organisms, I realised that it was an opportunity to put my personal experience and life-long studies of the desert into this book. I cannot claim that it is a complete text on the subject, but I hope that it becomes a useful addition to the scientific knowledge in this field.

Water is a paramount factor in determining the distribution of species on a global basis. The responses and adaptations of a species to water stress are critical for its success in any environmental niche. Numerous studies have reported a myriad of changes elicited by water stress. The changes observed are dependent on the species under study and on the severity, duration, and time course of the stress. The deserts of the Middle East provide a good example of the harsh conditions to which plants are subjected. This book is an attempt to elucidate the adaptations of these plants in order to sustain their life and reproduce to continue their survival.

It should be noted that the first field studies on the relationship to water of desert plants in the countries of the Middle East were carried out in Egypt (Stocker 1928a; Seybold 1929; Montasir and Migahid 1934; Migahid 1938; Migahid and Abdel Rahman 1953; Abdel Rahman and Batanouny 1964a,b, 1965a,b; Batanouny 1974). Many years ago, A.M. Migahid, Professor in the Department of Botany, Faculty of Science, Cairo University, established a school of botanists who were concerned with the study of the relationship to water of desert plants. The staff at the Hebrew University, Jerusalem, had begun studies in this area over 60 years ago (Evenari and Richter 1937; Oppenheimer 1951; Zohay 1962). Contributions to the study of the water economy of desert plants and the ecophysiology of these plants have been concentrated mainly in Egypt and Israel (Palestine). Successive generations of scientists in these two countries have continued this work. The author of the present book is among the former students of Professor A.M. Migahid. It is significant that the authors of two other books in the present series are among the former students of Prof. M. Evenari. After universities were established, in the 1960s to 1970s, in Arab countries of the Middle East, many Egyptian professors were appointed to them. The present author has been appointed at various times to Baghdad (Iraq), King Abdul Aziz (Saudi Arabia), and Qatar (Qatar) Universities. The studies of the staff of the Egyptian University have contributed to our knowledge of plant life in these countries. They have published books about the vegetation and plant life of this region, e.g. *Ecology and Flora of Qatar* (1981), and *Plant Life and Environment in Qatar* (1986a) by Batanouny, and *Flora of Saudi Arabia* by Migahid (1988–1989).

This book consists of 12 chapters. The first three chapters cover topics dealing with the deserts of the Middle East, climate aridity and topographic influence on plant life in the deserts. There is also an interesting chapter on the role of plants in the lives of the Bedouin. It is a well-known fact that Bedouin have an intimate relationship with their desert environment and its components, especially the plants. The Arabic heritage is full of information about and knowledge of desert plants, and the Bedouins in particular have a tremendous amount of indigenous knowledge of their native plants. Another chapter is devoted to discussing the concept of adaptation, its purposes, and causes. Three chapters describe the adaptations that promote increased water uptake and reduction of water output, as well as those which help the plants to tolerate and/or resist the drought. Escaping drought and adaptation to desiccation are two subjects treated in two chapters.

Salinity, being another major constraint to plant life in the deserts of the Middle East, is considered in a separate chapter. It discusses the adaptations and mechanisms by which halophytes can tolerate and/or resist salinity.

CHAPTER 2

Deserts of the Middle East

The term "Middle East" has no standard definition. During and after the World War II, it has been used to define a region covering all the Arab states of southwest Asia, including Turkey, Iran, Palestine and Cyprus. In many definitions Sudan, Egypt and Libya are also regarded as Middle Eastern countries; sometimes, Afghanistan is included as well. The constant use of "Middle East" to describe this region in communiqués and amongst military personnel during World War II made the term familiar to a large public. The present author feels uneasy at using to such a term, because he cannot help wondering east of what and middle of what? In this book the term is used to describe a vast area consisting of desert countries in that part of the world, i.e. many of the Arab countries, Israel and Iran (Fig. 1).

The area exhibits wide differences in its various regions as regards surface configuration, climate and vegetation. The climatic diversity is reflected in the vegetation, which ranges from dense humid forests to dreary plantless deserts. The plant life of the deserts of the Middle East is the main theme of this book.

Deserts cover a vast area of the Middle East. The arid and extreme arid zones make up by far the greater part of the region. These zones extend in a broad belt from the Sahara, in Libya, to the Red Sea and thence across the Arabian Peninsula into eastern Iran. Several countries including Libya, Egypt, Saudi Arabia, and the Gulf States are wholly arid. For example, in Libya about 92.5 % of the area receives an average annual rainfall less than 50 mm, and about 45.2 % receives less than 5 mm (Le Houérou 1982). In other countries, including Jordan, Iraq, Iran, the Yemen, and Oman, arid conditions cover the bulk of the territory. By contrast, Turkey, Lebanon and Cyprus have no arid areas.

One should take into consideration the conflicts and difficulties in defining what is meant by desert. Numerous investigators have proposed definitions of the word desert; however they are usually incomplete. Apart from this problem, the deserts of the Middle East are characterised by a considerable annual water deficit. From the botanical point of view, deserts could be defined as areas with sparse vegetation made up of special plants

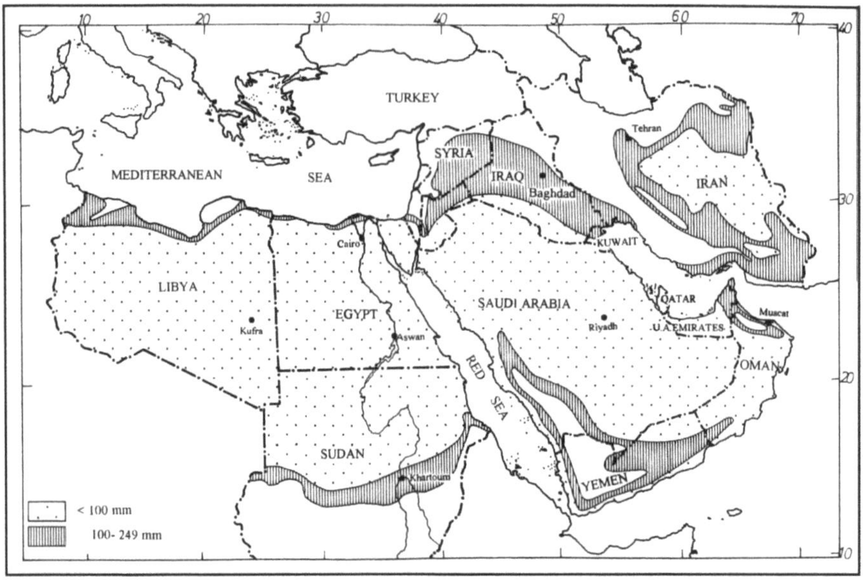

Fig. 1. A map of the desert countries of the Middle East showing areas with average annual rainfall less than 100 mm and between 100 and 240 mm

having particular characteristics that enable them to avoid, resist or tolerate drought. A paucity of trees is another common feature of deserts.

The deserts of the Middle East consist of both warm deserts, including the Sahara and the deserts of the Arabian Peninsula, and cold winter (continental) deserts like that of Iran.

Areas with an average annual rainfall below 250 mm are considered as deserts in the present treatise. However, distinctions have been made between areas receiving less than 100 mm/year and those receiving from 100 to 250 mm/year. The map (Fig. 1) shows that deserts receiving <100 mm/year occupy a vast area of the Middle East. However, this map does not illustrate the year to year variability of the rainfall, which becomes more marked as the average amount diminishes.

The most striking features of the deserts of the Middle East are two major sand deserts: the Great Sahara in Libya and Egypt, and the vast expanses of sand desert in the Arabian Peninsula, of which the largest is the Empty Quarter (*Rub' Al Khali*) in southern Arabia. The latter covers an area in excess of a half a million square kilometres. This is the largest sand desert in the Middle East. Both deserts are considered to be the harshest and the driest deserts on earth.

Stony desert plains and *hamadas* are common everywhere in the deserts under consideration. Further diversity is added by patches of salt marshes such as the Qattara depression in western Egypt. This depression has been excavated by wind erosion to 133 m below sea level. Salt deserts occur in basins with no drainage outlet. Water flowing in from surrounding areas in late winter and early spring lies on the surface to be evaporated during summer, leaving salt encrustations on the surface of the land and saline deposits in the soil profile. Such a phenomenon is very common in the oases in the southern desert of Libya and the Western Desert of Egypt, as well as the oases along the Arabian Gulf. The most extensive salt desert is found in the lower part of the *Dasht-e-Kavir* in Iran; similar conditions occur in the Qattara Depression in Egypt and the Dead Sea in Palestine (Israel) and Jordan.

Vast areas are occupied by coastal salt marshes because of the extensive shoreline, and numerous seas, lakes and gulfs are found in the study area. The presence of different soil types (saline, sandy and stony) has an effect on both the type of vegetation and the plant life in the various habitats.

It is extremely difficult to obtain reliable data about the actual areas of desert in the region. The boundaries of some countries are not well-defined and are indicated differently in various sources, including the statistical yearbooks of the particular countries, geographical texts, reports of UN agencies, and other regional organisations. Collated data on the areas of the countries concerned are given in Table l. Despite the fact that the total area of the Middle Eastern countries under consideration amounts to 12,440,130 km^2, only 679,700 km^2 (about 5.4%) are cultivated with field crops, vegetables and fruits. Irrigated agriculture is very limited in many of the countries of the region. For example, the area cultivated in the Arab Gulf states does not exceed 0.18% (Batanouny 1990). Only in Lebanon, Syria and the Gaza Strip (Palestine) is more than 30% of the land cultivated with crops are (Table 1). By contrast, the area occupied by rangelands (or, as given in some sources, as meadows and permanent pastures) is almost 3,958,620 km^2, or about 31% of the total area of the concerned countries. The area covered by forest and woodland is about 703,690 km^2 (about 5.6%).

The total land surface under use amounts to 5,341,510 km^2 (42%). The remainder (almost wasteland) covers 7,098,620 km^2. An extensive area is not used at all due to the environmentally harsh conditions, including the scarcity of water and the salinity of considerable tracts in the region.

It should be noted that the population of the Middle East is increasing rapidly. The estimated population of those countries with deserts was 253,567,000 in 1995 and was expected to rise to 479,910,000 by the year 2025. In view of the continuously growing population, shortages of food and other basic resources are ever increasing in these countries. This, inevitably, exerts a pressure on the environment and desertification is widespread. The estimates (better named "guesstimates") of the desertified areas reach 90% in the rangelands and rain-fed agricultural lands, and as much as 71% of

Table 1. Area, land use (×1000 km²) in 1991–1993 and population (×1000) in 1995 and calculated for 2025 in the countries with deserts in the Middle East. (World Resources 1997)

Country	Area	Crop land	Permanent pastures	Forest and wood land	Population 1995	Population 2025
African countries						
Egypt	1000	27.6	49.34	0.31	62931	97301
Libya	1759	21.76	133	8.4	5407	12885
Sudan	2376	129	1100	443	28098	58388
Asian countries						
Bahrain	0.66	3	6		843	693
Gaza and West Bank	6.2	45	x[a]	x	1481 (1976)	
Iran	1636	180.5	440	114	67283	123549
Iraq	475	54.5	40	1.92	20449	40565
Israel (pre-1967)	20.6	4.35	1.45	1.24	5629	7808
Jordan	88.9	4.04	7.91	0.7	5439	12039
Kuwait	17.8	0.05	1.37	-	1547	2805
Lebanon	10.23	3.06	0.1	28.19	3009	4424
Oman	212.4	0.62	10	x	2163	6094
Qatar	11.4	–	5	–	342	569
Saudi Arabia	2149	0.03	1200	18	17880	42651
Syria	183.7	57.7	801.8	67.9	14661	33505
United Arab Emirates	836	0.39	2	0.03	1904	2958
Yemen	527.9	148.1	160.65	20	14501	33676
Total	11310.79	679.7	3958.62	703.69	253567	479910

[a] x, Not available.

irrigated agricultural land in the Arab countries (Dregne and Chou 1992).This shows how severe desertification is in the region. The effects of humans on the desert vegetation may either be direct, i.e. acting on the vegetation cover itself, or indirect, occurring as a result of changes in other components of the ecosystem (Batanouny 1983). Human activities that affect the plant cover directly result in the deterioration of desert vegetation include the cutting or uprooting of ligneous species for firewood, overgrazing by farm animals, intensive collection of medicinal plants and removal of plants during the construction of roads, pipelines, etc. The human activities affecting the physical environment and thus indirectly leading to changes in desert vegetation include the introduction of irrigated agriculture, discharge of effluent and sewage, digging of sediments, shifting cultivation, and the activities accompanying development (building construction, tourism, off-road vehicles, war, military activities, etc.). Human activities and their severe impact endanger numerous species of desert plants. These plants are endowed with special traits and attributes that are the results of adaptations to the harsh desert conditions. Conservation of such plants is of utmost importance. The investigation of their adaptations to desert conditions is also important both from purely scientific and applied points of view.

The limits of the area under consideration are biogeographically rather unnatural, since they consist of rather small sectors pertaining to different floral regions. The Middle East is a meeting point of five main geographical regions: the Saharo-Arabian, Irano-Turanian, Sudanian, Mediterranean and Euro-Siberian regions (Zohary 1973). According to Zohary, the mere fact that five world-wide plant geographic regions meet here makes this region an area of special interest and raises a great many questions as to the past history of the respective floras, their advances and retreats, their centres of distribution and speciation, etc.

The meeting of different phytogeographical regions results in the presence together of different species from different regions – in other words, different species with different responses to the prevailing conditions of temperature and drought in the area. A community with different components and thus different requirements would be able to survive better than a single community having components with the same requirements. I do not refer here to chemical requirements, but mainly to water requirements or, better, the water-deficiency survival capacity. Moreover, plant life in the desert cannot simply be considered as a fulfillment of requirements, but as one of tolerance or adaptation in order to cope with the prevailing conditions or survival of drought. The heterogeneity of the components in the community leads to variations in the periods of time throughout the year when there is a high need for water. This can be seen as ecological "justice" – the distribution of resources by the distribution of activities.

The major part of the Middle East deserts lies within the boundaries of the Saharo-Arabian region. This is characterised by its mild temperatures in winter and high temperatures in summer. However, the main feature is the scanty, irregular rainfall. In many localities in this region, rainfall is not an annual recurrent phenomenon. The plant cover is usually of the accidental type in such habitats. A confined type of vegetation (Monod 1954; Walter 1963) is common in such deserts, i.e. except in sandy habitats, plant growth is mainly restricted to runnels and water-courses, where runoff water enriches the water revenue of these habitats.

The southern part of the desert region in the Middle East is affected by the conditions prevailing in the Sudanian region. These are characterised by summer rains and high winter temperatures. Trees such as *Acacia* spp. and *Ziziphus* spp. are able to penetrate to the north wherever warm winter conditions prevail. The same holds true with the mangrove *Avicennia marina*.

The northeastern part of the Middle East lies within the boundaries of the Irano-Turanian region, which is characterised by its continental climate and relatively low rainfall. It has two stress seasons, limiting plant growth: (1) a cold winter, when low temperatures are the limiting factor for plant growth and development, and (2) a hot and dry summer, when water is the limiting factor. Floral elements from this region penetrate into the northern part of the Saharo-Arabian region.

The Mediterranean region, although outside the desert proper, nonetheless has a marked influence on it, since its flora is well represented in the floras of various other territories (Zohary 1973; Orshan 1986).

CHAPTER 3

Climatic Aridity in the Deserts of the Middle East

3.1 Rainfall

The deserts in the countries concerned occupy the arid belt extending from North Africa to central Asia. The map in Fig. 1 (Chap. 2) shows the 100 and 249 mm isohyets (Blake et al. 1987). The area receiving less than 100 mm annually is vast and covers almost the whole of Egypt, Qatar, and Bahrain. Areas with rainfall between 100 and 250 mm occupy either the northern part of the Middle East or the extreme southern sector. Rainfall is characteristically scanty in most of the Middle East deserts. The average annual rainfall is shown in Table 2. A feature that is not evident from the map is the year-to-year variability in rainfall, which becomes more marked as the average diminishes. Several different methods can be used to measure this irregularity. One simple, commonly used measure is the quotient of variation, which is the ratio of the maximum annual rainfall to the minimum. This quotient is very high in the deserts of the Middle East. This can be deduced from the minimum and maximum values of annual rainfall at different meteorological stations in the Middle East (Table 2).

Irregularity is not only apparent from year to year, but there is also a wide variation between the corresponding months of different years. A few examples will illustrate this phenomenon. At Doha (Qatar, along the coast of the Arabian Gulf), the average rainfall in January is 19.8 mm, with a maximum of 101.8 mm in 1969; but during 6 years there was no rain at all during January (Batanouny 1981a). At Sidi Barrani (on the Mediterranean coast in Egypt) the average rainfall in January is 30.5 mm, but it reached 173.9 mm in 1964 and was only 24.2 mm in 1956 (Batanouny 1975). Similar fluctuations have been recorded at almost all the meteorological stations in the Middle East.

Among the longest records of reported rainfall are those at Muscat, Oman. Figure 2 shows the rainfall recorded at Muscat over the period 1893–1976. During this 85-year period, annual rainfall varied widely, ranging from about 10 mm in 1922 to 260 mm in 1944. The quotient of variation reaches 26 in Muscat.

Table 2. Temperature and rainfall particulars at different meteorological stations in the deserts of the Middle East

Station, elevation above mean sea level (m)	Temperature (°C)			Annual rainfall (mm)		
	Mean minimum of the coldest month	Mean maximum of the warmest month	Annual mean	Maximum	Minimum	Mean
Libya						
Bani Walid (230)	5.9	36.7	21.7	x[a]	x	62
Ghadams (326)	2.3	42.5	22.4	x	x	36
Kufra (301)	6.5	38.3	22.4	x	x	1.8
Sebha (440)	5.0	39.9	22.4	x	x	110
Sirt (20)	8.1	31.1	19.6	x	x	180
Egypt						
Alexandria (6.8)	9.3	30.4	20.2	313.6	33.2	192.1
Aswan (193.5)	9.5	42	25.9	x	0.0	1.4
Cairo (164)	8.6	35.4	20.4	63.4	1.5	23.8
Dakhla (110)	4.4	38.6	23.2	11.0	0.0	0.7
El Arish (15)	8.3	31	18.4	x	x	96.8
Kharga (72.8)	5.9	39.4	24.4	x	0.0	1.3
Mersa Matruh (30)	8.1	29.9	19.3	275	67.4	144.1
Quosseir (10.8)	13.8	33.8	24.3	x	0.0	3.4
Suez (3.1)	8.7	36.5	22.4	46.3	2.0	23.6
Sudan						
Karima (250)	11.9	43.8	29	x	x	25.0
Khartoum (380)	15.2	41.9	29.7	x	x	163
Port Sudan (5)	19.2	40.9	28.4	x	x	106
Wadi Halfa (190)	7.8	41.3	26.3	x	0.0	0.1
Israel						
Beersheba (270)	11	28	19.5	x	x	162.2
Eilat (10)	14.5	32.5	24.2	96.9	5.5	20.2

Table 2. (continued)

Station, elevation above mean sea level (m)	Temperature (°C) Mean minimum of the coldest month	Mean maximum of the warmest month	Annual mean	Annual rainfall (mm) Maximum	Minimum	Mean
Palestine						
Jericho	14.9	31.8	24.1	280	40	166
Jordan						
Aqaba (13)	16	32	25	x	x	24
Azraq (540)	x	x	18.6	x	x	76
Maan (1080)	7	24	17	x	x	48
Syria						
Albou Kamal	x	x	19.9	322.6	63.3	158
Athria (460)	1.3	36.5	17.6	x	x	196.7
Damascus (729)	2.4	35.7	17.5	284.8	97.8	220.9
Deir Ez Zor(203)	2.1	40	19.7	300.2	49.9	163.7
Palmyra (404)	2.3	37.8	18.5	285.5	38.6	133.0
Iraq						
Baghdad (34)	4	42	23	336	72.3	146.9
Baji (143)	7.6	34.1	x	363.0	41.0	182.5
Basra (4)	7	34	24.1	319.5	31.9	139.5
Diwaniya (20)	x	x	x	270.3	47.1	119
Habbaniyah (44)	4.3	43.7	22.7	200	46	124.9
Nasiriya (3)	x	31.6	24.0	221.1	27.8	111.5
Rutbah (616)	1.5	38	19	229.4	46.9	110
Shaiba (18)	5.9	35	25	237	38	133
Kuwait						
Kuwait (45)	7.7	44.7	25.6	242.4	23	115.1
Bahrain						
Manama (4)	14	38	x	169.4	10.1	76

Table 2. (continued)

Station, elevation above mean sea level (m)	Temperature (°C)			Annual rainfall (mm)		
	Mean minimum of the coldest month	Mean maximum of the warmest month	Annual mean	Maximum	Minimum	Mean
Qatar						
Doha (11)	12.7	41.2	26.5	302.8	19.7	78.1
Saudi Arabia						
Al Wajh (23)	8.8	41.2	24.3	x	x	22.6
Dhahran (17)	9.3	43.3	25.8	329.8	29.8	96
Jeddah (11)	14.5	41.6	28.2	173	17.6	79.8
Jizan (5)	19.5	40.6	30.2	x	x	30.5
Khamis Mushait (2066)	2.7	33.5	19	x	x	224.6
Mecca (240)	14.5	44.7	30.4	x	x	118.1
Medina (648)	6.5	45.5	27.9	103.8	0.7	47.2
Qaisumah 360	5.7	44.4	25.1	190.6	67.7	119.3
Riyadh (612)	8.1	43.2	26	257.7	21.8	90.1
Tabuk (774)	2.3	41.8	22.4	x	x	65.
Taif (1475)	6.5	37.3	22	x	x	220.9
United Arab Emirates						
Abu Dhabi (5)	13.8	29.4	27	176.9	2.4	61.5
Dubai (8)	13.7	29.3	26.9	220.6	9.0	90.7
Sharjah (2)	7.9	44.2	27.9	167.2	8.6	103.4
Oman						
Muscat (5)	16.5	40.9	28.3	260	10	103.7
Nizwa	8.6	41.7	25.7	x	x	198.
Salalah (18)	15.7	31.8	25.3	x	x	167.9
Sohar	12.0	38.9	25.3	x	x	106.3

Table 2. (continued)

Station, elevation above mean sea level (m)	Temperature (°C)			Annual rainfall (mm)		
	Mean minimum of the coldest month	Mean maximum of the warmest month	Annual mean	Maximum	Minimum	Mean
Yemen						
Aden (300)	25	33	29	93.0	7.6	39
Hodeida	x	x	30	x	x	89
Iran						
Abadan (3)	12	35	25	x	x	158
Bander Abbas (8)	18	33.5	26.7	x	x	147
Kerman (1650)	x	x	17.5	x	x	168
Tehran (1220)	2	30	16.5	x	x	246

[a] x, not available.

Irregularity and variability of rainfall are not only temporal, but also spatial. Consequently, rainfall is extremely unpredictable. Spatial variations in rainfall are evident when one investigates the annual rainfall at meteorological stations in a country as small as Qatar, where the maximum width of the country is 85 km, and the greatest length is 180 km. The amount of rainfall during the growing season, extending from September to August, exhibits considerable variation, as measured at seven reporting stations. Trainah, in southern Qatar, received 100 mm in 1973–1974; while Al Kharrarah, about 15 km away, received only 82 mm. During the same season, Dukhan and Ruwais received only 26.5 and 29 mm, respectively (Batanouny 1981a).

Rainfall measurements at Medina, Saudi Arabia, recorded over 11 years, show that the annual rainfall varied from 0.7 mm in 1973 to 103.8 mm in 1971. The quotient of variation reaches 148, which is among the highest recorded. The monthly rainfall at Medina (Fig. 3) also exhibits remarkable variation between the corresponding months in successive years. At Jeddah, records show that the annual rainfall varied from 17.6 mm in 1973 to 173 mm in 1968, while the average annual rainfall was 79.8 mm (Batanouny 1979a). Zohary (1973) stated that "it seems fairly clear that the means by which plants fight drought are not only effective against scantiness but also against irregularity of moisture supply".

Spatial variations in rainfall become even more significant when the monthly data are also considered. In February 1974, Trainah received 84 mm, while Al-Kharrarah received only 34 mm (Batanouny 1981a). The patchiness of the desert rainfall is a feature affecting to a large extent the distribution of plant growth.

An outstanding feature of desert rainfall is the occurrence of sudden, localised cloudbursts, which bring torrential rains. These sporadic storms are usually of the thunderstorm type. They have a paramount effect on both

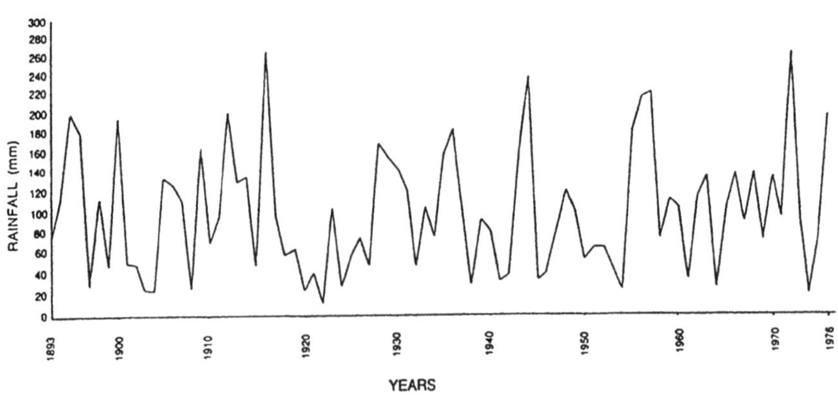

Fig. 2. Rainfall records at Muscat, Oman, during the period 1893–1976

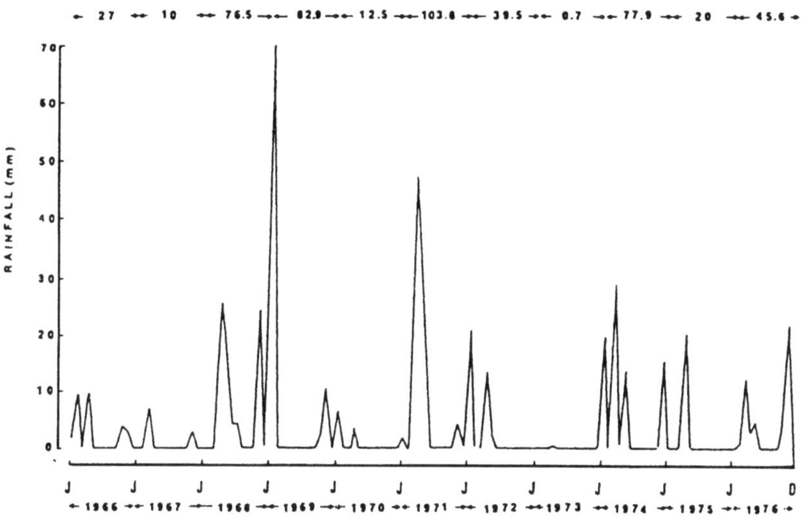

Fig. 3. Monthly rainfall at Medina, Saudi Arabia, over a period of 11 years (1966–1976)

the soil and on plant life. The wadis and water runnels are suddenly transferred into torrential streams sweeping the soil away and uprooting many plants. These ephemeral streams may also replenish the underground water reservoirs, which receive very little normal rainfall. The floods or *seyl* run down the valleys and destroy everything in their way. A few weeks after one such storm the ground surface in depressions is covered by a green mantle of plant growth, mainly of ephemerals. During *seyl* years, there is an ample chance for the establishment of new individuals of perennial species. This is due to the continuous wet profile extending from the surface down to the deep moist layers.

It is noteworthy that the rainfall on one day may exceed the average annual rainfall. For example, 70 mm of rain fell on 3 January 1969 at Medina (Batanouny and Baeshin 1983). Here, the average annual rainfall is 46.9 mm. At Jeddah, 88 mm fell on 17 April 1968, although the annual rainfall is only 79.8 mm (Batanouny 1979a). Such a phenomenon is common in the deserts of the Middle East.

A tragic flash flood occurred in June 1890, when a cyclonic storm hit Oman. Between midnight on the 4th and midnight on the 5th, 28.3 mm of rain fell. The water rose with such violence that 700 people were drowned. According to Arab historians, there have also been incidents of heavy spring rainfall that caused severe destruction to the villages.

The number of days with at least 1 mm rainfall is limited. In the deserts of the Middle East this occurs about 7 days/year. In areas with rainfall averaging more than 100 mm/year this number can reach 30 days.

The scanty rainfall and its spatial and temporal variability have a considerable effect on plant life. As a consequence, the soil moisture available to plants is very scarce and highly variable. The surface soil (surface deposits) is almost dry throughout the year except for a few days in winter and spring. Subsurface layers of deeper soils may comprise a permanently wet layer. This represents a permanent source of moisture for the deeply penetrating roots of perennials.

In Arabia, which represents the main bulk of the desert in the Middle East, the rainfall can be described in the terms used by the native Bedouin. They distinguish six seasons in the highlands of Saudi Arabia (Vesey-Fitzgerald 1957):

Gid: About 2 months, from early July to early August; this is usually a dry period, although there are often light afternoon showers along the watershed ridges.

Kharif (an Arabic word denoting autumn): About 1.5–2 months, from August to September, during which monsoon rains may come from the southeast, but they are irregular.

Wasm: Fifty days from early October to late November with heavy rains

Shita (an Arabic word denoting winter): About 3 months, from November to late February, with relatively heavy rains.

Smak: Fifty days from late February to mid April; this is usually a dry period.

Seif (an Arabic word denoting summer): From the last part of April to the beginning of June, with medium rainfall from the northeast.

These rainfall seasons are clearly defined in the highlands in southwest Arabia, where the effect of the monsoon is clear. Northwards, the rainfall occurs mainly in winter.

There is a prolonged dry season in the deserts of the Middle East. This extends in some areas to 8 or 9 months/year. The seasonal distribution of the rainfall varies from north to south. In the north, the rainy season extends from late autumn to early spring. Winter is during the rainy season. The dry season extends throughout the remainder of spring, summer and early autumn. By contrast, the southern part of the deserts of the Middle East, represented by the northern deserts of the Sudan, and the southern deserts of Arabia, has rainy summers. At Khartoum, the main rainy months are July, August and September. These are known as *kharif*. While the average annual rainfall in Khartoum is 163 mm, the rainfall in

July, August and September is 46, 75, and 25 mm, respectively. Negligible amounts fall in May, June, October and November. The difference in the seasonality of rainfall in the deserts of the Middle East is reflected in the kind of ephemeral species growing after the rains. In deserts with rainy summers, summer annuals dominate the ephemeral growth, while in areas with rainy winters, winter annuals represent the ephemeral cover.

3.2 Air Temperature

No doubt the air temperature is a limiting factor of paramount importance in the life of desert plants. The main characteristic of air temperature in the deserts of the Middle East is its wide variation, both daily and seasonal. The air temperature data given in Table 2 show that the plants in the deserts of the Middle East are subjected to both drastic temperature conditions during summer and wide seasonal and daily fluctuations in air temperature.

It should be noted that the temperatures given were measured with instruments placed inside standard meteorological screens, erected about 2 m above the ground surface. Therefore they do not indicate the temperature that exists on the ground surface or just above it, where plants grow. The wide amplitude of air temperature, both seasonally and daily, is even greater near the ground surface. This is an extremely important fact that influences the growth of plants in the desert.

Large areas of the desert in the Middle East, as well as their semi-arid fringes, experience extreme maximum air temperatures of over 50 °C. In accordance with standard meteorological practice such readings are "shade" temperatures in an instrument screen. There is rarely any shade in the desert and these readings are therefore less representative of actual temperatures experienced during the heat of the day than they might be elsewhere. Similarly, owing to the absence of vegetation and soil moisture in vast areas, the ground temperature is very much higher. At the other extreme, some desert places experience air temperatures below freezing point. Almost all parts of the desert have a clear seasonal rhythm of temperature, which is highest during the months of June to August and lowest during December to February. The wide amplitude of air temperatures, both seasonal and daily, evince the fact that plants, especially perennials, are subjected to severe fluctuations in air temperature. These facts can be seen in Figs. 4 and 5. Figure 4 shows air temperature records of different months at Doha airport, Qatar, on the Arabian Gulf. Figure 5 shows the hourly mean air temperatures in January and July 1977 at the same station. The mean hourly temperatures ranges from 13.1 °C at 04.00 h to 18.6 °C at 10.00 and 11.00 h in January. In July it ranges from 30.6 °C at 14.00 h to

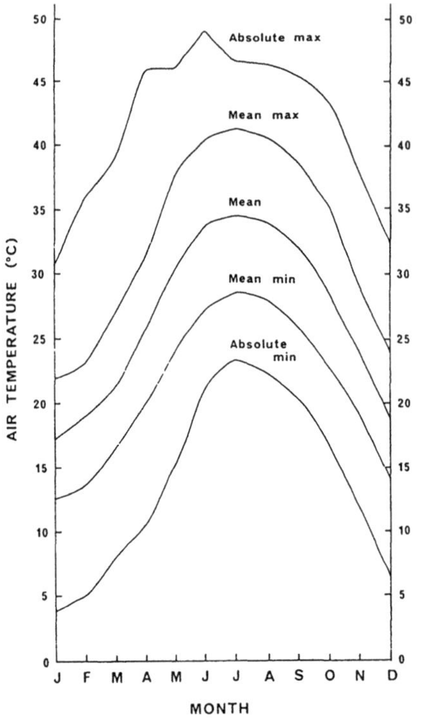

Fig. 4. Air temperature records of the different months at Doha airport (Qatar)

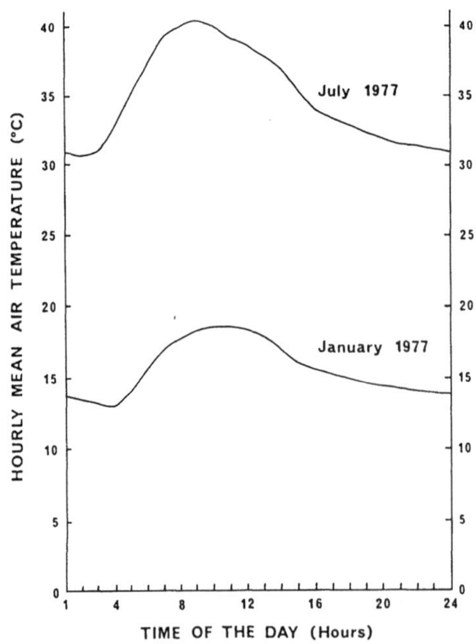

Fig. 5. Hourly mean air temperature in January and July 1977, Doha, Qatar

40.5 °C at 09.00 h. It is interesting that the maximum hourly means are reached before noon during both months.

The *Khamasin* in Egypt is a dry desert wind, which has many names in the countries of the Middle East. The wind is so dry that the relative humidity drops steeply, generally falling below 10 %. Temperatures are extremely high and may reach values between 40 and 50 °C. The physiological, in addition to the mechanical, effect on the plants is extreme, and the *Khamasin* presents a special challenge to all living organisms.

3.3 Air Humidity and Evaporation

Like air temperature, relative humidity undergoes wide daily as well as seasonal amplitude. On summer days, extremely low values are recorded during the early afternoon hours, while high values occur at night. The same pattern, but with a narrower range, occurs in the winter months. Relative humidity is not a statement of the actual moisture content of a volume of air, rather it is this amount as a percentage of the maximum quantity that air can hold at the prevailing temperature. Therefore, it must be remembered that the relative humidity is not of significant importance in understanding plant life without knowing the air temperature. Equal values of relative humidity do not mean that similar atmospheric humidity conditions exist, except when all other factors, including temperature, remain constant. For example, relative humidity values of 70 % can be recorded in summer and winter, but the amount of water vapour needed to saturate the atmosphere (100 % relative humidity) at low temperatures is less in winter than in summer. This is due to the fact that warm air can hold more water vapour than cold air. The capacity of air for invisible water vapour doubles with each increase of 11 °C in temperature. It follows that, when the relative humidity is 70 % at 20 °C in winter and at 31 °C in summer, the amount of water vapour needed to saturate the atmosphere (water vapour deficit) in summer is double that needed in winter even when relative humidity values are the same. Vapour pressure deficit (VPD) values are more significant in ecological studies than those of relative humidity and are usually high in desert areas.

In a study of the microclimatic variations among the different microhabitats in a desert wadi, Abdel Rahman and Batanouny (1966) observed that maximum values of the VPD were mostly attained at 13.00 h, but, in a few cases were attained at 10.00 or 16.00 h. Minimum values were recorded at night and in the early morning. The maximum value was recorded in October, reaching 55.9 mm Hg. The minimum, 0.3 mm Hg, was recorded in January. The same authors pointed out that the wadi bed has a higher VPD than found on the plateau, despite the relative richness of vegetation in the wadi bed compared with that on the plateau. This indicates the negligible effect of plant cover in the desert on the VPD. In his studies on the water balance of Egyptian desert plants, Stocker (1928a) came to the conclusion that "There is no case where a rise of relative humidity within the leafy frame-work of a desert plant has been proved; on the contrary in several instances, the humidity in the neighbourhood of transpiring leaves has shown a diminution. This phenomenon results from the fact that, on the one hand, the desert wind hinders any enrichment of the transpired water vapour about the transpiring organs, while on the other hand, the insulation reflected from the earth and also from the plants as heat favours an increase of temperature and a consequent lowering of relative humidity in proximity to the plants." It is interesting to note that this phenomenon was observed on several occasions by Cloudsley-Thompson (1962) in the Red Sea hills and the coastal plains, a major habitat of the desert locust.

The combination of high VPD, high temperature, maximum solar radiation and frequent moderate to strong surface wind creates conditions favourable for rapid evaporation if surface moisture is available. The evaporative power of the air, often referred to as the potential evaporation or potential evapotranspiration, is a difficult and rather theoretical concept in the absence of a permanent moisture source. Most evaporation measurements in the deserts of the Middle East have been made with Piche-type instruments mounted in an instrument screen but some measurements have also been taken from open water evaporating pans. Measurements of evaporation at various stations in the deserts of the Middle East range from low values in winter to high values in summer. Mean monthly evaporation values as high as 24.3 mm/day have been recorded in July at Dakhla, Egypt. At Abu Samra, in Qatar, open water evaporation reached 91 mm/month in December and 298 mm/month in June, with a yearly mean of 2,141 mm. The evapotranspiration values at the same station reached 85 mm/month in February and 344 mm/month in June, with a yearly mean of 2,244 mm. Months with high evaporation values are those without rain and with high temperature values. This is true in the areas with winter rainfall. It shows to what extent perennial plants are subjected to the severe evaporative power of the atmosphere during the period from April to September, or even longer.

3.4 Types of Climate

There are several different systems for the classification of climate, and numerous investigators have tried to make use of them. None of these systems takes into account the variability of rainfall. This is very important

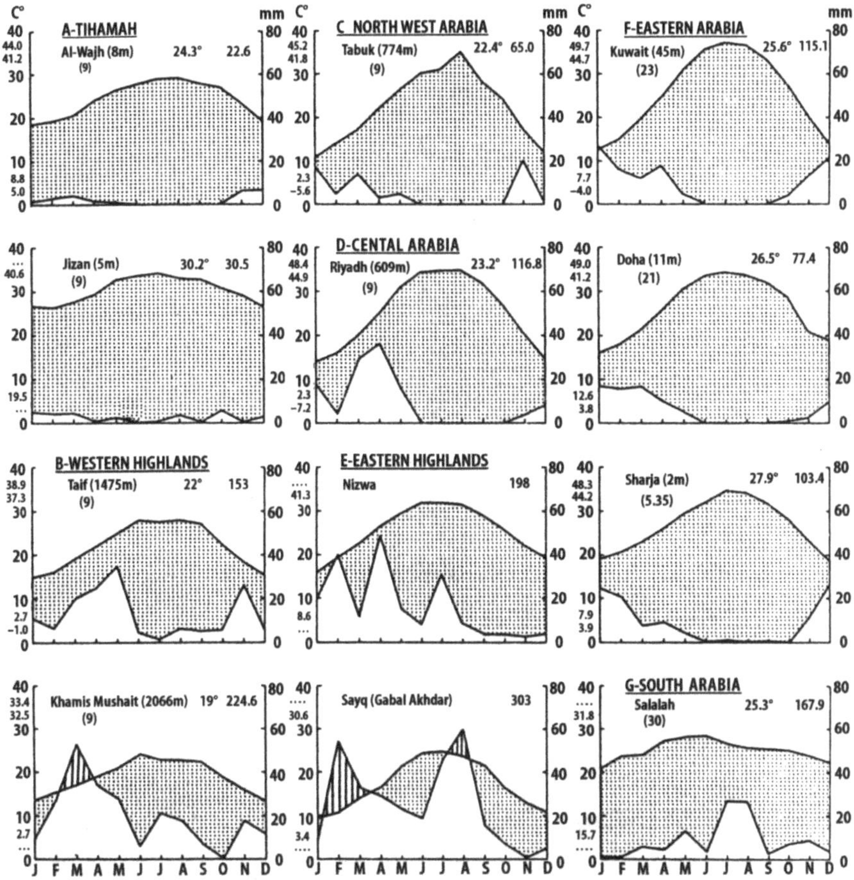

Fig. 6. Climatic diagrams of some meteorological stations in the Arab Gulf countries

in studying deserts where rainfall is extremely irregular and unpredictable.

In general, the climate in vast areas of desert in the Middle East is characterised by extreme aridity. Winter is the rainy season in the northern part of these deserts. In the southern part, rainfall occurs in summer.

Walter (1955a, 1957; Walter and Lieth 1959), by means of "climatic diagrams", suggested a method of expressing climatic aridity. Figure 6 shows that at many of the representative meteorological stations on the Arabian Peninsula precipitation curves underlie the temperature curves throughout the year. According to Walter's climatic diagrams of Africa and Asia (Walter and Lieth 1959)), the climate of vast areas of the deserts of the Middle East is of the extremely arid type. At a few stations, the area between the temperature and precipitation curves is reduced, indicating less aridity. Whenever the precipitation curve overlies the temperature curve, it means a humid period, but this phenomenon appears only at stations in the highlands of south Arabia.

CHAPTER 4

Topography and Plant Life

Zohary (1973) pointed out that in the Middle East "here plant life and death rate are very tightly correlated with the topography and the shape of the landform"; that is, one might state that the water resources supply is controlled by topography and local relief The rainfall in many parts of the deserts under consideration is not sufficient to support any permanent vegetation. However, thanks to the irregular topography, relatively dense vegetative cover can be observed in depressions, wadis and runnels. Lower areas may receive a relatively considerable supply of water (Figs. 7, 8), while at the same time, at higher elevations most of the rain that is received runs down to the lower regions. As a consequence, higher regions lose much of their water supply and surface deposits. Therefore, two or more different habitats, each with its particular environmental conditions, which in turn are a result of topographical irregularities, can be observed in neighbouring sites.

Walter (1963) wrote: "If the runoff in a region with 25 mm precipitation is 80 %, and the area in which the water accumulates constitutes 4 % of the total area, that area will receive an amount of water corresponding to a rainfall of approximately 500 mm." This shows how much water the main channels of the wadi can receive. Where there are no wadis, the accumulation of sheet runoff in depressions is of great importance in enriching water revenues.

The types of vegetation coincide with the pattern of the water resources of the habitat. This can be considered to be a communal adaptation to the changing environmental conditions. Three main types can be distinguished: (1) the *accidental* (Kassas and Batanouny 1984)type is found in those areas where rainfall is not an annually recurring incident, e.g. in the Great Sahara and the *Rub'Al-Khali.* (2)The *restricted* type (Walter 1963), or mode contractée (Monod 1954), occurs in arid areas where rainfall, though low and variable, is an annually recurring phenomenon. This type is the result of runoff and the accumulation of water at lower elevations. The vegetation is confined to rather restricted areas (wadis, runnels, and depressions with relatively adequate water supply). (3) The *diffuse* type results in vegetation that is more or less evenly distributed (rainfall desert;

Fig. 7. A wide runnel receiving runoff water from the adjacent high hills after a heavy rainstorm in North Qatar

Zohary 1962). This type occurs in areas with considerable rainfall (above 100 mm per annum).

It is not only the water resources of a particular habitat that are controlled by the local topography but also the physical and chemical attributes of the soil (Batanouny 1973a), and the soil thickness and its texture are evidently affected by it. A slight depression of a few centimetres in the ground level will lead to the accumulation of a thin veneer of soil carried with the water running into it. Hence, a relatively suitable habitat for plant growth is created despite the harsh conditions of the desert (Batanouny and Hilli 1973). Tracks of cars, even many years old, e.g. from World War II, are usually inhabited by vegetation denser or less dense than the surrounding sites depending on the type of soil.

Variation in topography creates different microhabitats, which differ considerably in their characteristics, though they lie only a few metres apart from each other (Batanouny 1973a). The microhabitats in desert wadis differ from each other with respect to their soil attributes (Batanouny 1973a), water supply (Abdel Rahman and Batanouny 1965a,b), microclimatic conditions (Abdel Rahman and Batanouny 1966), and consequently vegetation (Abdel Rahman and Batanouny 1965 c).

Fig. 8. Low regions in the local topography receive considerable amounts of runoff water causing the growth of relatively dense perennial vegetation

In one, the wadi Hoff in the Egypto-Arabian desert, irregularities in topography have led to the creation of a number of well-defined microhabitats: (a) plateau, (b) shaded microhabitat, (c) wadi bed proper, (d) first terrace, (e) second terrace and (f) slope (Fig. 9). There are always at least as many microclimates as there are types of obstacles. The factors that induce the segregation of microclimates in the horizontal plane are the macro- and microtopography, the qualitative constitution of the substratum, and the variation in the structure of the vegetation cover.

The open plateau affords the least chance for plants to appear, except in slightly depressed parts in which sand has accumulated. The wadi bed proper is usually devoid of plant cover, due to the destructive action of floods. The soil conditions in this microhabitat only allow for the appearance of a few ephemerals. Owing to its relatively deep soil with a deeply seated, permanently wet layer, the first terrace of the wadi is the most favourable microhabitat for perennials as their deep roots have a continuous water supply. In the shaded microhabitat, the accumulation of fine material, which is sheltered and protected from erosion, creates relatively the best soil and moisture conditions. These are very favourable for the development of climax vegetation. In the second terrace the vegetation is scarce and poor in number

Fig. 9. Wadi Hoff; a wadi in the limestone Egypto-Arabian desert showing various microhabitats with different plant cover due to the topographic irregularities

of species due to the extremely low moisture content and high salt content of the soil. The slope is covered by rock fragments and is devoid of plants except in pockets filled with fine soil where a few ephemerals can grow.

The vegetation in all the microhabitats is always open, a characteristic that has an important value in water economy. The plant cover amounts to 10, 20, 20 and 3 % in the plateau, shaded microhabitat, first terrace and second terrace, respectively. The existence of open vegetation with widely spaced perennial plants reduces root competition, which, in turn, minimises the amount of water withdrawal by roots per unit volume of soil. Even after the development of annuals, reduction in competition between perennial roots still holds, due to layering of the roots. The reduction in root competition also favours the free development of roots, which is one of the most important factors in drought resistance.

It is interesting to note that shading due to local topography creates particular shaded habitats harbouring special plant species adapted to the prevailing conditions. Even species that grow in sunny and shaded habitats exhibit remarkable flexibility and plasticity in their responses to the environmental conditions in each of these habitats. This difference in responses correlates with the different vapour pressure deficit (VPD) values recorded in each of the two habitats (Abdel Rahman and Batanouny 1966).

Fig. 10. Sheep and goats grazing along a meandering runnel in the Summan Desert, northern Saudi Arabia

The mosaic pattern of desert vegetation provides a picture of the underlying irregularity in the factors affecting plant life in the desert, the most important of which are the water resources. These are mainly, especially on a local basis, the result of topographic irregularities and their consequences. As a result of the variation in the physical environment at any particular site, the biological components of the site are changed. Different individuals of the same species also exhibit different responses to the un-even micro-environmental conditions. This, of course, indicates plasticity in adaptation, especially of the annuals growing in depressions with varying soil thicknesses from their bottoms to their edges. So, plant life exhibits differences with respect to both the different sites and within the same site. This has a paramount influence on grazing animals. Figure 10 shows the role of topography on the distribution of plant life in the Summan Desert, northern Saudi Arabia, and consequently on the path of the grazing animals within a limited area in the desert.

Within the area defined as desert in the present treatise, there is a remarkable variation in altitude, from −390 m below sea level to almost 2500 m above sea level. Such wide altitudinal variation is reflected in the climate, soil properties and, consequently, plant life.

CHAPTER 5

Plants in the Arabic Heritage

The Middle East is unique as regards its plants and the available traditional information about them. Perhaps it is the only area on earth in which one can trace the presence of some plant species since time immemorial. Plants and plant products are frequently mentioned in the Bible and the Quora'n (Batanouny 1992a). Numerous plants have been discussed in the Sayings (*Hadith*) of the Prophet Mohammed (Batanouny 1986b). Classic Arabic poetry contains innumerable references to desert plants and even to their habitats, phenology and morphological features. This is due to the vital importance of plants in the lives of Bedouin.

The Middle East is the cradle of famous and very old civilisations, e.g. the Pharaonic, Assyrian and the Babylonian. Plants and plant names were depicted on the walls of the temples and on papyri.

The deserts of the region have always supported a hardy race of herdsmen. For the sake of survival, the Bedouin have had to be acquainted with the names of the plants, their growth season and phenological aspects and their palatability or toxicity. Nomadic pasturalism enables the Bedouin to dominate vast areas and traverse hundreds of miles with their flocks to reach the verdant patches occurring at various periods of the year at various sites. The Bedouin, from their childhood onwards, live in intimate relation with the plants growing in their environment. They have an understanding and a sense of biological diversity.

In classical Arabic, there has long been a clear distinction between *khullah* (sweet plants), the glycophytes of today and *hamdh* (salt plants) or halophytes. Examples of the *hamdh* (halophytes) were provided by Moslem philologers, lexicographers and writers, e.g. *shnan* (*Seidlitzia rosmarinus*) *harm* (*Zygophyllum* sp.), *girm* or *shurah* (*Avicennia marina*), and *ikrish* (*Aeloropus lagopoides*).

In old Arabic books and classic Arabic dictionaries, there has often been reference to ecotypic variations, such as the differences between the desert ecotypes of *Lycium* and *Ziziphus* and the ecotypes living in moist habitats. It was pointed out that the latter types have fewer spines.

It is noteworthy that in their classical writings the Arabs gave specific names to so-called community types, i.e. defining vegetative cover and its habitat together. These names give an accurate picture of the prevailing habitat features of the community in which a particular plant dominates. The phytogeographic distribution of plants was also considered in the writings of the Arab scholars. They could distinguish between the native homes of the different poets by the names of the plants mentioned in their poems. If a poet mentions *salam* (*Acacia ehrenbergiana*) and *samr* (*Acacia tortilis*, Fig. 11) he is from Hejaz, but when he mentions *arta* (*Calligonum comosum*) then he is from Najd. Such information can be obtained from classical Arabic books written more than 1000 years ago.

For many centuries, social organisation was based on the tribal system. Thus, life in the desert was governed by division of the herding population into groups and subgroups, thereby maintaining a balance between grazing resources and animals. Effective rangeland management within these societies was achieved with the *hema* (an Arabic word denoting protection; Draz 1979; Batanouny 1984). It consisted of different types of protection with special systems of ownership or use. As a result of recent socioeconomic and

Fig. 11. *Acacia tortilis* (Arabic: *Samr*) growing in a *rodah* (depression) in Qatar. The plant is common in the deserts of Arabia and is cut for firewood and grazed by camels

political changes, the *hema* system has been abandoned and the rangeland subjected to free and communal grazing. Consequently, the degradation of the rangeland is notable everywhere in Arabia.

Before the establishment of modern states on the Arabian peninsula, the principal socioeconomic activities of each Bedouin (stock raising, agriculture, settling) could be carried out within the *dirah*, or territory of his tribe. Usually, the tribal *dirah* was large enough to provide all the resources necessary for the yearly cycle of transhumance. The activity of the Bedouin was limited by customary rules (*orf*), and protected by retaliation. During periods of drought, which affected the grazing area of a single *dirah*, threatened groups were often allowed into the territory of a wealthier tribe, to which they were united by a pact of alliance (*mohalfah*). This usually occurred on the basis of negotiations of an invariably reciprocal character. Besides limiting access to pasture, tribal norms curbed freedoms in the use of water. Over the years there was thus created a general organisation, based on alliances and agreements, that permitted rational exploitation of resources as variable as those of rangelands. In addition to solving social and economical problems, this organisation introduced essential regulations for range management, since it was intended to control stocking rate on the basis of the available forage produced in any year. The *hema* (Pl. *ahmyah*) system evolved within the framework of the social organisation of the tribes. It was a grazing reserve belonging to a tribal group who had the exclusive right of exploiting it. This system helped to protect range plants from extinction through misuse and communal overgrazing. No doubt, it preserved valuable genetic resources, which we are unfortunately overexploiting today.

Range reserves with concepts similar to the Arabian *hema* are recorded in other countries: the *mahmia* or *marah* in Syria, the *koze* on the Syrian-Turkish-Iraqi borders (*koze* is the Kurdish word for *hema* in this region), *ghidal* and *zeneika* in North Africa.

Draz (1979) gives the following types of *ahmyah*:

- Animal grazing is prohibited. Cutting of grasses is, however, permissible during specific periods and during droughts. A specified number of members of each family are allowed to cut mature grass during the season.
- Grazing and/or cutting is permitted, but is restricted to certain seasons of the year.
- Grazing is allowed all year-round. The kind and number of animals permitted for grazing are, however, specified. The reserve is kept for beekeeping. There is a limited number of these *ahmyah* and grazing restrictions are relaxed once the flowering season is over.
- The reserve aims to protect forest trees.

A unique type of *hema* existed close to Damascus (Syria) for about five centuries up until 1930. According to government documents, a 100 ha area was maintained as pasture land for aged or unfit horses until the end of their lives (Draz 1980).

Today, *ahmyah* survive almost uniquely in isolated mountain communities in Saudi Arabia. Some *ahmyah* are endowed with written statutes, while others do not have such documents and are handed over through heredity on the basis of de facto situations. Ownership deeds (*sanad*) stipulate the permitted exploitation methods of the members of the clan and the sanctions given to strangers who trespass, or who have rights but do not respect the regulations.

The exploitation norms vary in different *ahmyah*. In a large *hema* in Bilad Zahran, south of Taif, Saudi Arabia, direct grazing is forbidden, but not until the hay has grown to a full arms length can it be mown; so, if drought stunts the grasses, haymaking is postponed. Adult male members of the clan are allowed to cut the hay on one day a week and the women on another day. Those who are authorised may cut from dawn to dusk, and tribal guards (*hurras*) watch their work. These guards supervise the *hema* and depend on a sort of "administrative council", whose members are called *muwamin* and who are elected for life by the male members of the clan. The *muwamin* from time to time decide if and when hay can be cut, what maintenance is to be done (border markings, paths, etc.), whose turn it is to stand watch, and how trespassers should be penalised.

The norms that prevailed in the old institutions in the Arabian peninsula have been greatly suppressed in recent years. The opening of all rangelands to free grazing has resulted in denudation of the plant cover, serious erosion and frequent destructive floods. The largest country on the peninsula, Saudi Arabia, has changed from a livestock exporting to a livestock-importing nation. This also holds true for the other oil-producing countries in the region.

In traditional Arab dictionaries, lexicographers give innumerable plant names with descriptions of their life forms, geographical distribution and uses. Among the most famous of these dictionaries is *Al Mokhassas* (Ibn Sidah, 1007–1066 A.D.), *Lisan Al Arab* (Ibn Mandhour, d. 1311 A.D.) and *Qamoos Al-Muhit* (Fairuzabadi, 1329–1415 A.D.). There is also a rich pharmacological and botanical literature by Arab scholars, e. g. Ibn Al Bitar (d. 1248 A.D.), who travelled in North Africa and the Near East collecting plants and information about them. In his book, *Histoire de la medicine Arabe*, Leclerc calls him "the greatest botanist of the East". Ibn Al Bitar wrote a great pharmacological and botanical encyclopaedia. One of the more famous Arab botanists is Rashid Ad-Din Ibn As Suri (1177–1243 A.D.) who lived in Syria and travelled in the Near East accompanied by a painter. He not only described many unknown plants, but also had them painted both fresh and preserved. Unhappily, his book has been lost, but mention of it can be found in many dictionaries and other sources.

Before ending this concise note on the role of the Arabs in botanical studies of desert plants, one should mention that the Arabs had names for particular habitats in the desert. These names have been latinised and are still used by modern scientists. Names of Arabic origin include *hamada, reg, serir, sabkha,* and *wadi.*

Many Latin names for plants growing in the Middle East are derived from their original Arabic names (Batanouny 1985). As examples we might mention *Zilla,* from *silla,* an Arabic word denoting a spine; *Retama,* from the Arabic *retem* and the Hebrew *rothem; Jasminum,* from *yasmin; Cadaba,* from *cadhabah; Rokama,* from *roqamah; Cuminum,* from *cammoo;* and the specific epithet *termis,* from *termis.* The specific epithet *pyrotechnica* was given to *Leptadenia,* as Forsskal noticed that the Arabs were using the fibres of its stem to produce fire using stone and flint.

Plants were valued as a natural resource by the Bedouin. They even gave the names of plants to themselves. Men and women were named after plants such as: *arfaj* (*Rhanterium*), *handhal* (*Citrullus colocythis,* Fig. 12), *salama* (*Acacia ehrenbergiana*), *samr* (*A. tortilis,* Fig. 11), *murr* (*Centaurea*), *sudairy* from *sidr* (*Ziziphus*), *suwaidy* from *Suaeid* (*Suaeda*) and *morikhy* from *Markh* (*Leptadenia*).

In the ninth century A.D., Arabs began recording their heritage of poetry, philology, religion and medicine, in addition to translating a large number of works from Greek and other languages. This resulted in the appearance of an enormous number of books and treatises dealing with botanical nomenclature and different aspects of plant life and uses. To give a few examples, one may mention Al Asma'i (who died about 831 A.D.) and his *Book of Plants.* Some useful botanical material was included in the book of Abu Zaid Al Ansari (d. 830 A.D.), which was written and published by one of his disciples, Ibn Khalawih (d. 980 A.D.). The famous book of Abu Hanifa Ad-Dinawary (d. 895 A.D.), *The Book of Plants,* discusses the names of plants, their habitats and their life forms (trees, shrubs or forbs). Its thoroughness and the care taken in the description of each species mark this work, which combines a philological, historical and botanical approach to the study of plants. It has been widely read by numerous other authors and cited on many occasions over the centuries. It includes numerous quotations from poetry and philological interpretations of the verses.

In the fourth century (AH) (10th A.D.) several philosophical studies of plants appeared. Nasr (1976) writes: "The Ikhwan al-Safa devoted one of their Epistles to the morphology, genesis and manner of growth of plants as well as the numerical symbolism of their various parts and their place in the total cosmic order". The *Kitab al-I'tibar* (The Eastern Key) of Abd al-Latif al-Baghdadi is particularly rich in its description of the plants of Egypt.

Muslim studies on botany deal mostly with topics such as the classification of plants, their physiology, genesis and modes of growth, description of their parts, their relationship to geographical and climatic conditions and

Fig. 12. A fruiting branch of *Citrullus colocynthis* (Arabic: *Handhal*); a desert creeping perennial plant

their medical as well as "occult" properties (Nasr 1976). It is interesting to mention that these and other books classified plants (of course, desert plants) according to very efficient criteria: e.g., the presence or absence of spines, the colour of the flower, the shape of the fruit and the habitat features and geographical distribution were cited. The spiny plants were further divided into herbs, shrubs and trees.

CHAPTER 6

Adaptation of Desert Plants

In this chapter the ability of desert plants to cope with their environment will be discussed. The traits that enable these plants to function (e.g. germinate, grow, reproduce) can be described as an *adaptation*; that is, certain plants are endowed with *adaptations* that allow them to live under conditions that would be very hostile for other groups of plants.

When desert plants are transferred to the Arctic, they are unable to survive there. This shows that adaptations are specific to particular situations and special conditions. Every trait is developed as the result of a certain constraint. The life of desert plants is thus a matter of tolerance and adaptability. The traits that help desert plants to tolerate arid conditions are true adaptations. The function of a given trait in terms of its arising through the process of natural selection can be seen as its evolutionary "purpose" (Amundson and Lauder 1994). The adaptations of an organism – its form, physiology, and behaviour patterns – cannot easily be separated from the environment in which it lives. In other words, adaptations of organisms and the development of their environments go hand in hand.

Before investigating the various adaptations that have been recorded in desert plants, it is important to more extensively define the term adaptation and to discuss the causes of adaptations.

6.1 What Is Adaptation?

Adaptation means the process by which an organism becomes modified, by a change in structure, form or function, such that it is better able to survive under given set of environmental conditions. It is the result of a process of change. To adapt means to change and to become suitable to meet new needs and respond to different conditions. Amundson (1996) stated that etymology has been kind to the term adaptation in its evolutionary context. He referred to the Oxford English Dictionary (OED), which traces the root "apt" to the Latin "apt-us", meaning "fitted, suited, appropriate". It is derived from "ap-ere", to fasten, attach – for example, as applied to a garment. Both the verb "adapt" and the noun "adaptation" are from French and Late Latin, with the first cited English uses in 1610 and 1611 (Amundson 1996). In the OED, adaptation is defined as "1) The action or process of adapting, fitting or suiting one thing to another. 2) The process of modifying a thing so as to suit new conditions. 3) The condition or state of being adapted. 4) A special instance of adapting." These definitions reveal that the term adaptation can mean the process of adaptation itself, as well as the result of this process. Amundson (1996) suggested that the process of adaptation gives rise to a *generic state* of overall adaptedness and to *specific traits,* which are adaptive modifications and are for this reason called adaptations. He defined the generic state as the relationship between an organism and (roughly) the environment to which it is adapted, while an adaptation (adaptation *qua* individual trait) is the modified part of an organism that performs a particular biological function for that organism and thus contributes to its state of adaptation.

Theoretical biologists have engaged in discussions of the relationship between the process of adaptation and an actual trait designated as an adaptation (Gould and Vrba 1982). Amundson (1990) proposed that a characteristic should be called an "adaptation" only when the intention is to assert that the characteristic has evolved to serve a specific purpose. One might refer to the statement of Sober (1984) when he analysed the problem as follows: "*A* is an adaptation for task *T* in population *P* and only *A* became prevalent in *P* because there was selection for *A*, where the selective advantage of *A* was due to the fact that *A* help perform task *T*."

Adaptation has been described as a state of equilibrium (Reeve and Sherman 1993). It is represented by the traits that help in the performance of a specific function. Amundson and Lauder (1994) suggested that for a trait to be an adaptation (historically defined) it must have a precise function. They wrote: "describing an organic trait as an adaptation has meant either: (1) that it benefits the organism in its present environment (whatever the trait's causal origin), or (2) that it arose via natural selection to perform the action that now benefits the organism."

In the present book, we will be looking at the kinds of phenomena that can be seen to be adaptations to harsh desert conditions. What can be regarded as the cause of the adaptations is obvious: deficiency of water, heat stress, salinity, and so on. Such an idea leads us to consider whether or not these conditions are constraints. The term "constraint" implies some sort of restriction on variety or on change. It is well known that environmental peculiarities may have drastic effects on plant life through their impact on various physiological processes within the plant body. In this context, constraints are seen as enhancing the possibilities for adaptive changes (Amundson 1994).

Acclimation is different from adaptation. The term acclimation means that the organisms become accustomed to new environments or new conditions. When an organism is introduced to a new environment and proves to be successful there, it is said to be acclimated or acclimatised. Acclimation or acclimatisation is a positive response to new conditions. It is not adaptability, as adaptations are genetically fixed traits that are transferred from generation to generation, as can be seen in the case of what are called ecotypes. Acclimation can be achieved through phenotypic responses and/or physiological responses that are not transferred to subsequent generations. However, it should be noted that the ability of an organism to acclimate to new conditions is not unlimited. Moreover, acclimation may persist for only a short period of the organism's lifetime.

6.2 Why Adaptation?

The majority of plants are rooted in the ground and cannot escape from the conditions into which they have been born. They are therefore continuously exposed to the full impact of extreme external conditions and consequently must adapt to them (Evenari et al. 1971). The desert environment imposes severe stresses on the plants growing in it. These stresses include:

1. Water stress due to the low, variable and irregular rainfall and to the high evaporative power of the atmosphere and low atmospheric humidity.
2. Heat stress due to high temperatures (in the hot deserts of the Middle East) during summer, or frost stress (in the cold deserts of Iran), as well as wide diurnal and seasonal changes in the air and soil surface temperatures.
3. High solar radiation on most of the days of the year.
4. High wind velocity causing burial of the shoot systems of plants or eroding the soil and thereby exposing the root system. Such exposure can also be due to the infrequent torrents from thunderstorms. The velocity of dry wind greatly increases the evaporative power of the air, thus impairing the water balance of plants through its effect on transpiration.
5. Salinity, and in particular saline habitats, both coastal and inland.

Desert plants, especially perennials, grow under the impact of the combined effect of the above-mentioned stresses. The main problem confronting these plants is water balance. In view of the water deficit in the soil, they would be expected to have acquired several traits helping them to absorb water under desert conditions. At the same time, these plants live under a severe deficit of atmospheric humidity, which causes excessive water loss through transpiration. This indicates that the desert plants must be endowed with characteristic adaptations and traits that help them to carry out a variety of different processes.

Individual desert plants show an extraordinary flexibility, both in their structure and in their physiologic-metabolic processes, that reflects the extreme seasonal temperatures and moisture changes of the environment. The mechanisms of drought resistance and/or tolerance in one species may not be manifested in another, even one growing under the same environmental conditions. Every species has its own potentialities for enduring or resisting drought. Migahid et al. (1972) stated that the efficiency of the mechanisms of drought resistance varies with changes in the environmental conditions. Micro-variation in climatic or edaphic conditions leads to the suppression or enhancement of the effectiveness of one or more mechanisms. Evenari et al. (1971) wrote "the more extreme the environmental conditions are, the more specific and pronounced the adaptations, the greater their variety among the plants in one habitat, and the greater the individual variability of adaptive responses to the environment. As a result, desert plants show an extraordinary individual flexibility of structures as well as a flexibility of the physiologic-metabolic processes that reflect the extreme seasonal temperature and moisture changes of the environment."

One can speak of the diversity of adaptations as a reflection of the genetic diversity within the different individuals of any population. This "physiological diversity" can be seen as a response to the prevailing environmental conditions and their reflection in the genetic attributes. This physiological, though genetically fixed, property exhibits considerable plasticity and flexibility. The flexibility of the traits and adaptations of desert plants is itself an adaptation and is emphasised in this group of plants as a response to the unpredictability of environmental conditions and to the micro-environmental variations. The mosaic pattern of the distribution of plants in the desert is a reflection of this diversity, both in the environment and in the adaptive responses of the plants.

Zohary (1973) rightly stated that drought resistance, as generally understood by physiologists, is less characteristic of desert plants than is drought survival, and leads to the preservation of the desert flora no matter the cost to the plant.

Ecologists usually regard the terms resistance and tolerance as interchangeable. Though both have more or less the same function, the words have different meanings. Resistance can be thought of as withstanding the impact of particular conditions and, at the same time, remaining unharmed. The plant's ability to do so arises from the adaptations acquired during its evolution. In other words, resistance arises from genetically fixed traits or adaptations. On the other hand, tolerance means the quality of being able to suffer the hardships of the severe conditions without being damaged. It is to be noted that this is a negative response. The tolerance of a plant comes from its characteristics and responses. To explain both terms, one can imagine a plant under conditions of drought. This plant resists drought by modifications and adaptations, and thus tolerates the effects of this drought. Flexibility of resistance and tolerance mechanisms is an important feature of desert plants, which may suffer from constraints due to drought, but nevertheless tolerate these hardships. In the meantime, they resist these hardships with astonishing flexibility, showing adaptations and modification of different traits.

6.3 What to Adapt to?

Adaptations and acquired characters should fulfil one or more aspects of the following:

1. Increased absorption of water, to at least compensate for the water lost by the plant in its attempts to maintain a positive water balance. This is achieved by adaptations of the root system, as well as by the high osmotic pressure of the cell sap.
2. Control and reduction of the water lost during transpiration. This is achieved by:
   a) Reduction of the transpiring surface, or at least of the exposed surface, e.g. by leaf orientation or rolling. Orientation of the leaves is a measure by which leafy plants reduce exposure to direct solar radiation.
   b) Reduction of the transpiration rate under dry conditions, both daily and seasonal. Adaptations relating to the control of transpiration are important. In particular the evolution of stoma, which allow a plant's transpiration rate to be related to its water status, has been among the most important evolutionary steps in successful colonisation of the land (Newman 1976).
   c) Communal reduction of water loss; either by the disappearance of individual plants or by a decrease in the green foliage and biomass of the transpiring surface. The simultaneous occurrence of different life forms with different water requirements within the community assures an ecologically balanced distribution of water, as well as of water output of the community, all year-round.
3. Tolerance and/or resistance to drought through the biochemical characteristics of the cytoplasm or by the production of amino acids that protect the protoplasm. Tolerance can be partially achieved through the particular anatomical features of desert plants. Desiccation tolerance by

resurrection plants, desert lichens and algae, is an extreme example (see Chap. 11, this Vol.).

4. Evasion and escape from drought. This occurs either through the complete drying out or death of above-ground parts of perennial geophytes in summer, or the disappearance of annual plants with the advent of the dry season. Partial death for the sake of survival is a common feature of perennial desert plants.

CHAPTER 7

Adaptations That Increase Water Absorption

The ability of any plant to keep its tissues turgid depends on a balance between the external relationships of water uptake and water loss, as well as on the internal osmoregulation of the plant cells. Water uptake from the soil, especially in deserts, is a crucial function of every root system and largely determines the water status of the shoot. The most obvious starting point for research in such a case is to examine whether drought-tolerant plants have any special means of obtaining water that are denied to drought-sensitive species. The root systems of desert perennials have particular characteristics and adaptations that favour water absorption from as large a volume of soil as possible.

In addition to its function in water uptake, the root can communicate changes in the soil environment to the shoot (Freitas 1997). Freitas (1997) stated that roots seem to be the primary sensors of water deficits, and biochemical signals transported from the roots cause many of the physiological perturbations observed in the shoots of plants subjected to drought. Freitas (1997) also states that these signals are probably related to changes in the kind and concentration of hormones exported to the shoots, possibly in the transpiration stream.

7.1 Adaptations of the Root System

The importance of depth and extension as well as other characteristics of the root system in enabling the plants to develop under the adverse moisture conditions that prevail in the desert has been emphasised by numerous investigators (Migahid and Abdel Rahman 1953; Hammouda 1954; Lemée 1954; Kausch 1955; Abdel Rahman and Batanouny 1959d; Oppenheimer 1960; Migahid 1962; Zohary 1962; Evenari et al. 1971; Batanouny and Abdel Wahab 1973; Batanouny 1980).

From the anatomical point of view, Fahn (1989) listed the following properties of the root systems of plants as adaptations to desert conditions:

1. The form of the root system
2. The succulence of the roots
3. Development of thick bark
4. Sclerification of the cortical cells
5. Isolation of the vascular cylinder by periderm formation or by necrosis of the cortical parenchyma

Desert plants cope with limited water availability and high temperature by exhibiting various root strategies (Nobel 1996). The root systems of desert plants possess various characteristics, which represent adaptations to the prevailing dry conditions in the desert. The presence of such characteristics and adaptations helps the plants to absorb water in quantities at least sufficient to compensate for that lost during transpiration. In other words, to keep the water balance positive in the plant bodies.

The main characteristics of the root systems of desert perennials that act as adaptations increasing water intake are described in the following sections.

7.2 Characteristics of the Root Systems of Desert Plants

7.2.1 Rapid Growth in the Seedling Stage

Seedlings establishing themselves in semi-arid and arid regions are highly susceptible to mortality from water stress. The survival and growth of seedlings in xeric conditions have been shown to be adversely affected by competition, both from other seedlings and from adults, and it is generally assumed that such competition is for soil water (Abdel Rahman and Batanouny 1959d; Fowler 1986; Reichenberger and Pyke 1990).

The capacity of the root system for rapid and early development is an important factor in drought resistance (Aamodt and Johnston 1936; Abdel Rahman and Batanouny 1959d). Lösch (1989) reported that, with decreasing water availability, root growth is enhanced at the cost of the accumulation of above-ground biomass. The results obtained by Reader et al. (1992) confirm that the rooting depth of seedlings may increase significantly in response to drought. However, this is not the case with all plant species. This same group of investigators stated that plasticity in seedling rooting development may

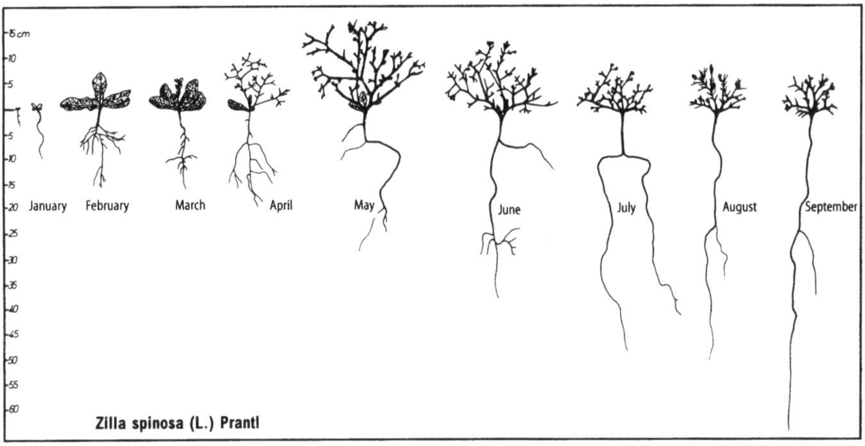

Fig. 13. Root systems of *Zilla spinosa* individuals over the period from January (seedling stage) to September (establishment). Note the rapid growth of the root system during the growing season. (After Batanouny 1960)

differ among species for different reasons. Under desert conditions, the germinating seedlings of perennials fail to establish themselves when their roots do not succeed in reaching a permanently wet layer before the onset of the dry season. Abdel Rahman and Batanouny (1959d) found that the roots of the newly germinating seedlings of *Zilla spinosa* exhibit continuous growth at a moderate rate till the end of the wet season. By the beginning of the dry season, the roots of an experimental seedling exhibited considerable growth (Fig. 13). The root system was only 18 cm deep before April, but showed progressive growth, reaching 65 cm in August. However, it showed very limited growth in September near the end of the dry season. By such means, perennial plants escape progressive desiccation of the upper soil layers and thus succeed in establishing themselves. Bhardwaj and Gopal (1979) found that root elongation of the perennial herb *Tephrosia apollinea*, which grows also in the deserts of the Middle East, is much greater than shoot growth. They found that the young seedlings of the 2–3 leaf stage with a shoot less than 1 cm long may have roots up to 30 cm or more in length. Slightly older seedlings have roots up to 50 cm long. It is to be noted that the roots of some ephemerals in the same area in the Eastern Desert of Egypt did not penetrate deeper than 21 cm (Figs. 14–16). In years with very low rainfall, a thin layer of the surface soil is moistened, leaving a dry horizon between the wet surface layer and the permanently wet layer at a greater depth. This intervening dry layer acts as a barrier, limiting the downward extension of roots. The root growth of some ephemerals and a perennial (*Zilla spinosa*) are shown in Fig. 17. The remarkable difference between the growth rates of ephemerals and those of perennials can be seen from this figure. It shows the

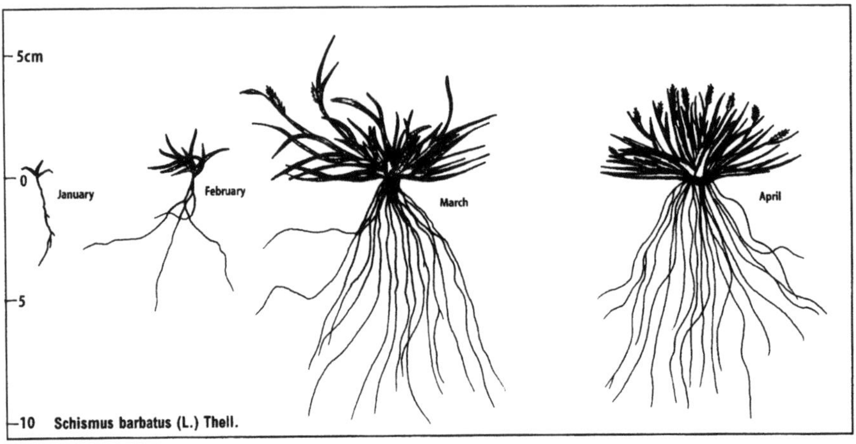

Fig. 14. Development of the root system of *Schismus barbatus*, an annual desert grass, from the seedling stage in January to the fruiting stage in April. (After Batanouny 1960)

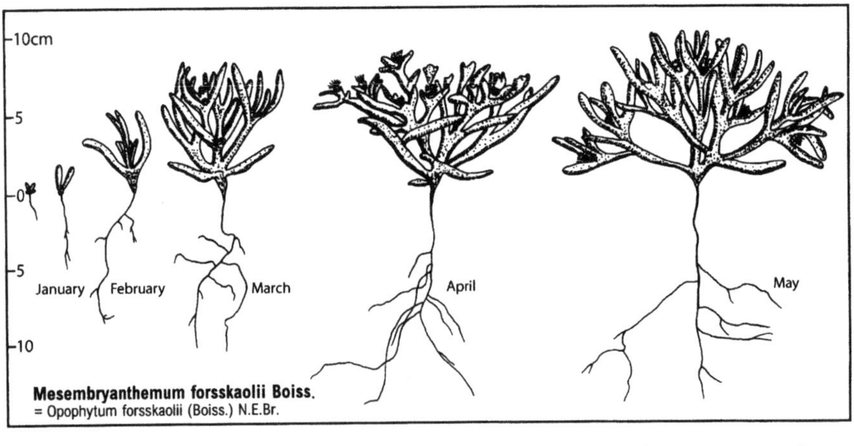

Fig. 15. Development of the root system of *Mesembryanthemum forsskaolii* (syn. *Opophytum forsskaolii*; a succulent ephemeral) from the seedling stage in January to the fruiting stage in May. (After Batanouny 1960)

rapid growth of the roots of perennial plant seedlings, a characteristic of utmost importance to the life history of desert plants.

Another case in which rapid root growth occurs is that of desert geophytes with shallow roots. The bulbs or rhizomes of geophytes, which are buried in the soil and remaining dormant during the dry season, are well protected by layers of dead tissue or cork from drying out and overheating (Evenari et al. 1971). It was reported that shortly after the first rains and as soon as the soil in the vicinity of the bulbs or rhizomes is wet, they form roots and rootlets with incredible speed (Evenari et al. 1971). The authors also

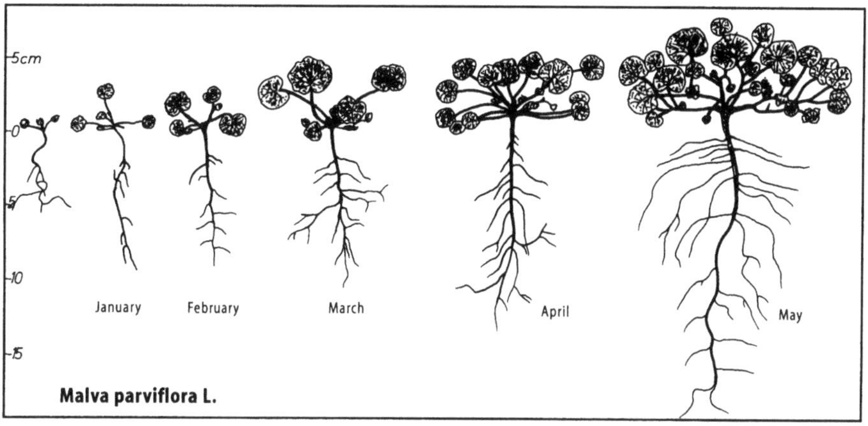

Fig. 16. Development of the root system of *Malva parviflora* (an ephemeral herb) from the seedling stage in January to the fruiting stage in May. (After Batanouny 1960)

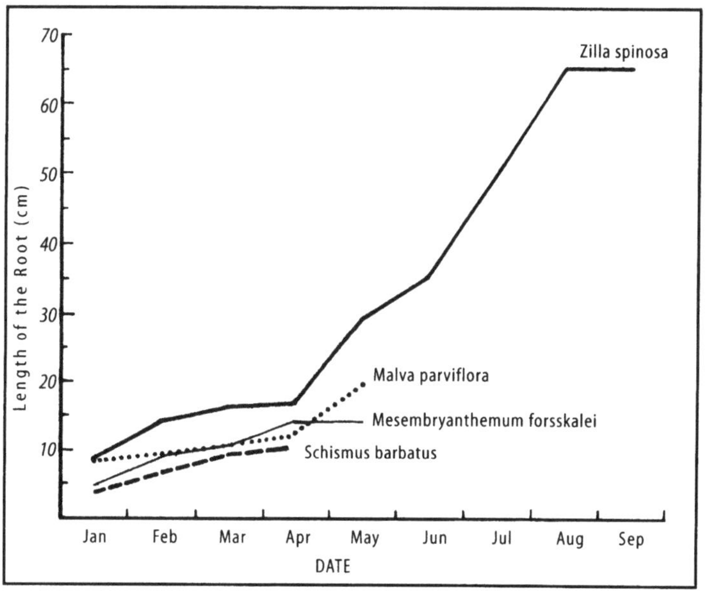

Fig. 17. Growth of the root system of the desert perennial *Zilla spinosa* compared with that of the ephemerals *Schismus barbatus, Mesembryanthemum forsskalei* and *Malva parviflora.* (After Batanouny 1960)

found that, less than 12 h after the first rain had wetted the upper 2 cm of soil, the first rootlets appeared on the short-styled sedge (*Cavex pachystylis*) and a few hours later their dense network enmeshed the upper soil layer.

7.2.2
Root Extension

It is well known that increased efficiency of moisture absorption is primarily due to the development of an extensive root system. This is a common feature of desert perennials (Kausch 1968; Batanouny and Abdel Wahab 1973; Kassas and Batanouny 1984). Studies in the deserts of the Middle East have revealed that perennial desert plants have extensive root systems (Abdel Rahman and Batanouny 1959d; Zohary 1962; Kausch 1968; Evenari et al. 1971; Batanouny and Abdel Wahab 1973).

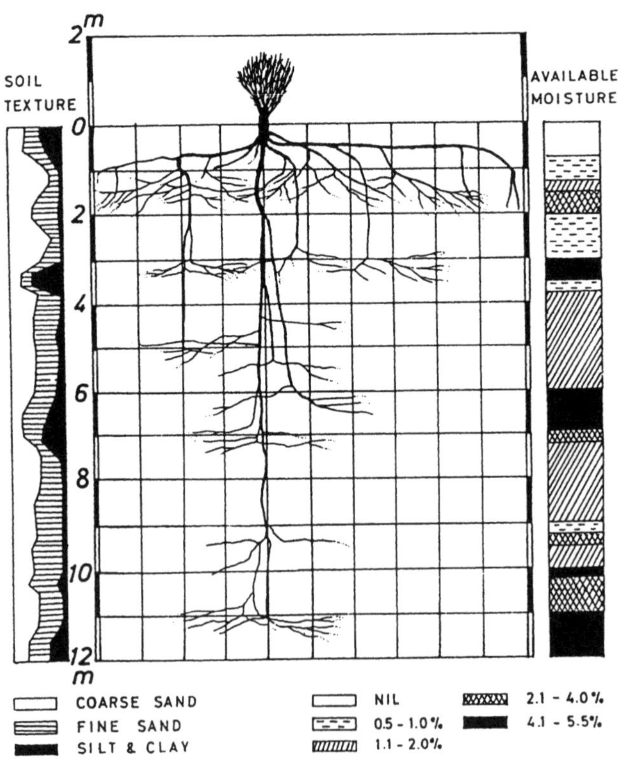

Fig. 18. Distribution of the root system of *Leptadenia pyrotechnica* (Forssk.) Decne with respect to texture and the percentage of available moisture in different soil strata down to 12 m. (After Batanouny and Abdel Wahab 1973)

Root extension consists of different features including: (a) deep penetration of the roots, enabling plants to absorb water from deep permanent wet layers; (b) root extension to pockets in the soil between stones, which helps in providing the plant with available water (Kausch 1968); and (c) lateral extension, both extensive and intensive, to exploit large soil volumes. All these features ensure a high root/shoot ratio and represent adaptations that increase water uptake by the plants.

Under desert conditions, the superficial soil layers are almost dry throughout the year, except on a few days during the rainy season. This is due to evaporation and the absorption of water by shallow-rooted ephemerals. However, these plants represent dry mulch, protecting the deeper layers from further evaporation. This helps in maintaining the soil moisture content of the deep layers at levels allowing the roots of perennials to absorb water from them. This deep layer that contains moisture, usually at depths below 50 cm, is called the permanent wet layer (Migahid and Abdel Rahman 1953). Roots of the shrub *Leptadenia pyrotechnica*, which is 160 cm high, were found to extend to a depth of 11.50 m with the diameter of the lateral circular extension being 10 m (Fig. 18). This root system would exploit about 850 m^3 of soil with a total available moisture of 23,000 kg. The annual water output from this bush reaches 5700 kg. This means that the available water in the soil occupied by such a root system would be sufficient to supply the plant for a period of 4 years without replenishment by rainfall.

In the eastern desert of Egypt, Kausch (1968) found that the root system of *Deverra tortuosa* (syn. *Pituranthos tortuosus*) reaches a depth of 5 m, while Zayed (1984) found that the root system of *Retama raetam* extends to a depth of more than 2 m. Even old, suberised roots absorb considerable amounts of water (Kramer 1946).

7.2.3
Low Shoot/Root Ratio

It is a common feature among the desert perennials that these plants develop root systems that are larger than their shoots. Doubtless, this is an important factor in survival. The moisture collected by the extensive root system is drawn upon for the consumption of a reduced shoot. The preponderance of the root system over the above-ground shoot system facilitates adjustment of the water balance of desert plants. The transpiration rate of desert plants is very high when calculated on the basis of the fresh weight or the area of transpiring shoot, but it is very low when calculated on the basis of the fresh weight or the area of the absorbing root (Batanouny 1980). Each unit of transpiring surface is supplied with water by many more roots than is usual in mesophytes. Hence, a positive water balance is more likely to prevail. In other words, the low shoot:root (S:R) ratio must be advantageous to ensure an adequate supply of water, which is the major limiting resource in deserts. The low S:R ratio is established as early as the seedling stage. The ratio of shoot height to root axial length (length of the root excluding lateral branches) is 1:7 in *Leptadenia pyrotechnica* (Fig. 14).

Tilman (1988) postulated that, due to the physical separation of above- and below-ground resources, plants face an unavoidable trade-off in their abilities to compete for these resources. In order to obtain a larger portion of one resource, plants must allocate more biomass to structures involved in the acquisition of that resource at the expense of allocation of biomass to structures involved in the acquisition of another resource. Morphological plasticity may increase the competitive ability of a plant over a range of different resource availabilities (Grime et al. 1986; Crick and Grime 1987; Tilman 1988; Aerts et al. 1991).

Reynolds (1986) stated that an investment of energy into root growth is necessary for obtaining water and the essential nutrients required for growth and photosynthesis. Gadgil and Gadgil (1979) stated that the S:R ratio is not a result of competition between these systems but reflects a balance of function that ultimately maximises carbon gain and, hence, success in a given environment.

7.2.4 Stratification

Different species in a plant community exploit different layers above and below ground. Ephemerals differ from perennials regarding the strata exploited by their roots. Some plants that form mounds exploit the soil water resources from the body of the mound above the ground surface. This results in a distribution of the layers that are exhausted.

Uneven water penetration results in a heterogeneity of the water contents of the soil. This is reflected in the distribution of root systems within the soil profile.

The root systems of some of the plants growing on the coastal dunes along the Mediterranean Sea exploit different strata. These plants have roots that penetrate deeply into the soil in addition to roots extending horizontally and parallel to the soil surface. The deep vertical roots extract water from the deep layers, whereas the horizontal roots make use of the dew that condenses during one of the 180 nights during the year.

The occurrence of roots of the same or different species in the various strata of the soil provides an opportunity for reducing competition for water and for obtaining water from various strata. Specific differences in the distribution of the root system exploit the lack of homogeneity of the water in such soils and thereby reduce spatial competition (Rundel and Nobel 1991).

7.2.5
Sparsity of Root Systems Within the Soil: Absence of Root Competition

Desert vegetation. is characterised by its openness, which results in the absence of competition between perennials, which represent the permanent framework of the vegetation. The withdrawal of moisture from the soil depends on the density, cover and openness of the vegetation (Abdel Rahman and Batanouny 1965a,b). The sparsity of roots per unit volume of soil reduces root competition which, in turn, minimises the amount of water withdrawn by the root per unit volume of soil (Figs. 19, 20). These characteristics ensure less withdrawal and better conservation of the soil moisture. The dry season lasts, in some places, for up to 9 months. During this time the uppermost soil layers are dry and form a protective layer that eliminates water loss from the soil through evaporation. Water loss from the soil in the dry season takes place only through absorption by roots and its subsequent transpiration (Abdel Rahman and Batanouny 1965b).

Ephemerals exploit the excess water that the perennials are unable to utilise in years of good rain. If, as in years of drought, there is no such surplus, the ephemerals simply do not appear.

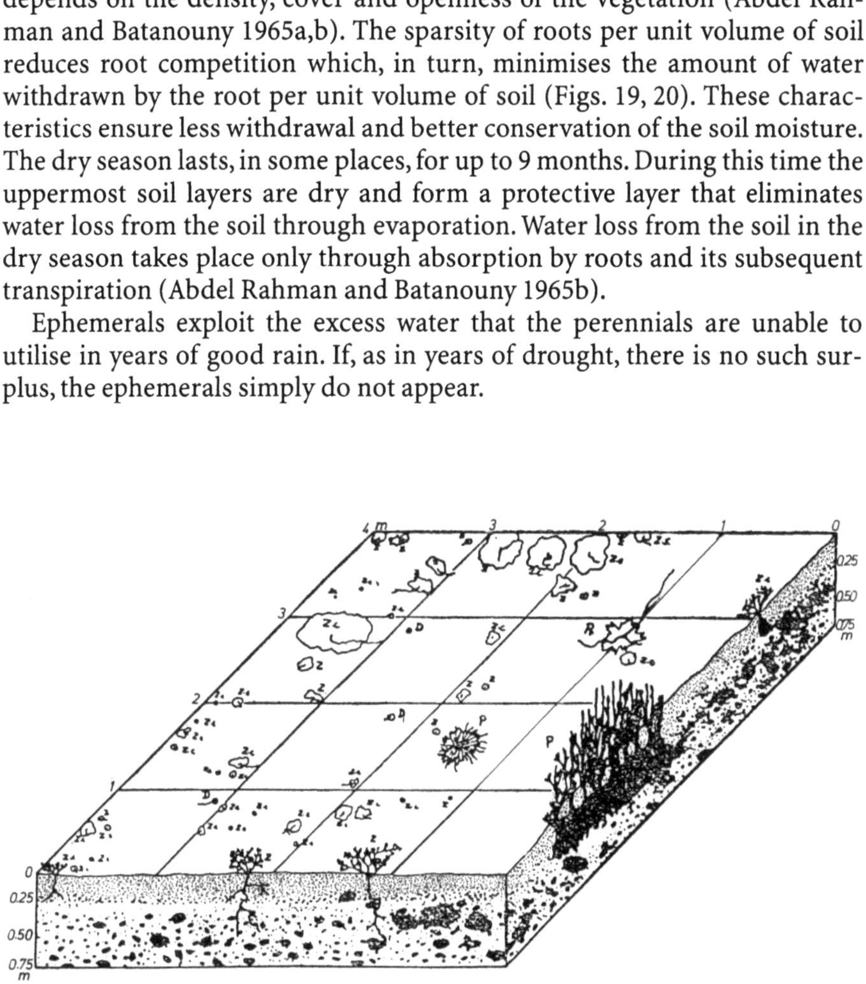

Fig. 19. Horizontal and vertical distribution of roots in an area 4 × 4 m on a terrace with relatively dense vegetation along the main channel of wadi Hoff, Egypt. (After Abdel Rahman and Batanouny 1965 c)

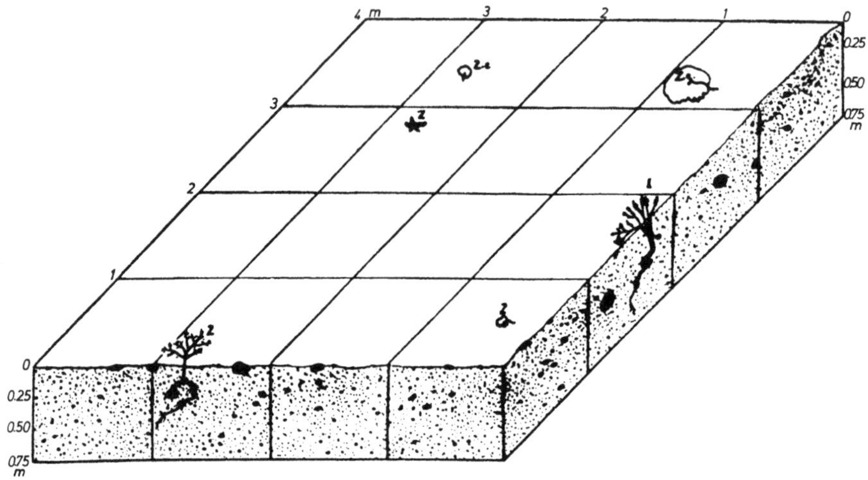

Fig. 20. Horizontal and vertical distribution of roots in an area 4 × 4 m on the plateau with sparse vegetation, wadi Hoff, Egypt. (After Abdel Rahman and Batanouny 1965 c)

7.2.6
Plasticity of Responses to Soil Conditions

Morphological plasticity facilitates the growth of plants in heterogeneous environments. Halophytes growing in sites with a shallow water table have their root systems confined to the layers above it. In sites where the water table is far below the surface, these plants have deep roots. When the soil is stony the rooting habits of desert plants are influenced accordingly. The layer of soil below large embedded stones is protected from evaporation and the water content is consequently higher than in other parts of the same horizon. This leads to the formation of dense tufts of thin roots, especially in the wet season, when the surface layers are enriched with the permeating water. The formation of these tufts adds greatly to the absorbing surface.

Soil strata with different physical attributes have very different water-holding capacities and support growth of different types of roots. In zones with high water-holding capacity the stored water engenders the ramification of roots at a considerable rate (see Fig. 18).

The roots of desert shrubs and dwarf shrubs are specially adapted to the moisture conditions in their habitats. Evenari et al. (1971) stated that the root systems of shrubs growing on hilltops, slopes, or loess plains are very shallow and do not penetrate to more than 30–50 cm, since deeper soil is lacking or the deeper layers are permanently dry. The lateral extension of

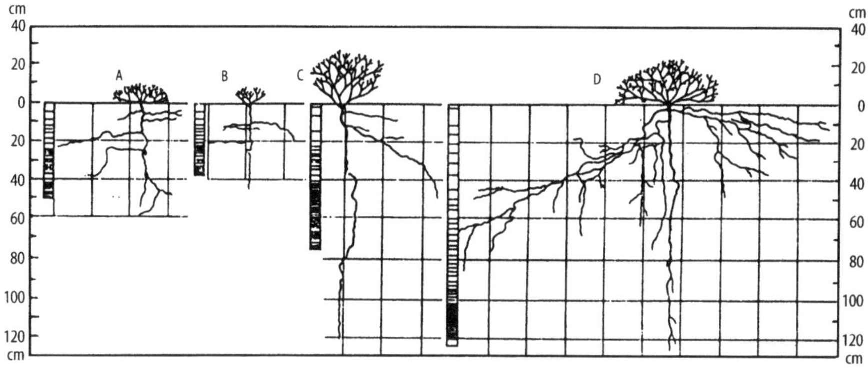

Fig. 21. Root systems of *Deverra tortuosa* (syn. *Pithuranthos tortuosus*) inhabiting different soils supporting various plant associations. Penetrability is represented by histograms (the wider the lines, the more penetrable the soil) shown beside each root system (according to Zaghloul 1928). Associations of *A*, *Salsoa tetrandra*; *B*, *Anabasis articulata*; *C*, *Artemisia herba-alba*; D, *Plantago albicans* and *Echiochilon fruticosum*. (After Batanouny and Zaki 1969)

the root system, however, becomes enormous, and out of all proportion to the size of the shoots.

Fitter (1991) claimed that the plasticity of root development is an adaptive variable, and Grime (1979) suggested that plasticity may be a key attribute of a plant's adaptive strategy. Shallow-rooting species with less plasticity in rooting depth have less opportunity to escape when the soil dries than do deep-rooting species (Cook 1943; Osonubi and Davis 1981; Reader et al. 1992).

In species with wide ecological amplitude, the root system of the same species exhibits considerable plasticity according to how it is affected by the soil conditions. For instance, Batanouny and Zaki (1969) studied the root systems of *Thymelaea hirsuta* and *Deverra tortuosa* (syn. *Pithuranthos tortuosus*) growing in soils with different attributes and supporting different plant assemblages (Fig. 21). They concluded that, despite the powerful influence of inheritance on the type of root system of the species, the penetration of the root and the number of laterals are modified according to soil conditions. The penetrability of the soil was among the main factors affecting the depth of root penetration, lateral extension, and formation of laterals. This indicates that the plasticity of the root system is a response to the soil conditions, especially those related to water supply.

7.2.7
Production of Adventitious Roots in Hillocks (Shoot-Borne Roots)

The burial of the stems of some desert plant as a result of the formation of phytogenic hillocks usually leads to the production of adventitious roots in the body of the mound. These roots differ in their shapes (Batanouny and Batanouny 1968, 1969). The formation of these phytogenic hillocks and their role in the rooting of plants are discussed below.

7.2.7.1
Formation of Phytogenic Hillocks

Dry open desert regions and sandy beaches with scanty vegetation are often swept by violent winds. Obstacles that check the velocity of the wind may cause some of its load to be deposited, and mounds or ridges are formed. The gradual building-up of these mounds is usually a physical process only partly due to plant growth, as wind-blown sand, dust and water-borne sediments may accumulate around the plant which then traps the sediments.

Numerous desert and marsh plants are able to build mounds around their vegetative growth (Fig. 22). Batanouny and Batanouny (1968, 1969) described these plants as follows:

1. They are able to produce adventitious roots from vegetative organs that become buried under the collected sediments. These roots may be long and thread-like, travelling horizontally at a shallow depth over long distances, as in *Arthrocnemum macrostachyum*. They may be in the form of tufts of very fine roots as in *Anabasis articulata*. Tufts of fine adventitious roots cover the thick buried stems. These roots may be responsible in great measure for the total water absorption. Figure 19 shows part of *Stipagrostis scoparia* with adventitious roots produced at different levels. Production of these newly formed roots is enhanced by the accumulation of sand.
2. They are capable of producing new shoots to replace those buried by on-blowing sand. In this way, plant growth copes with the sand accumulation: plant growth keeps abreast of the sand inundation.
3. They usually have intricately branching shoot systems, thus effectively intercepting wind-borne and water-borne materials and causing their deposition around the plant body.
4. They have perennating organs buried in the soil, or shoots and buds close to or near the soil surface. In other words, they have definite life forms:
   (a) Geophytes with rhizomes, e.g. *Ammophila arenaria, Sporobolus virginicus, Pennisetum divisum, Sprobolus spicatus* and *Juncus acutus*.

Fig. 22. The mounds formed by *Limoniastrum monopetalum*. North coastal zone, Egypt

(b) Hemicryptophytes, especially *Hemicryptophyta caespitosa* or the tussock-forming species, e.g. *Stipagrostis plumosa, S. ciliata, Lasiurus sindicus (*syn. *L. hirsutus* and *Frankenia revoluta.*

(c) Chamaephytes:

(i) *Chamaephyta pulvinata* (cushion plants), e.g. *Halocnemum strobilaceum, Arthrocnemum macrostachyum* (syn. *A. glaucum*), *Suaeda vermiculata, S. fruticosa, S. pruinosa, Salsola tetrandra, Haloxylon scoparium* (syn. *H. articulatum*), *Anabasis articulata, A. setifera, Cornulaca monacantha, Zygophyllrim album* and *Limoniastrum monopetalum.*

(ii) *Chamaephyta graminida* (hard grasses), e.g. *Lygeum spartum.*

(iii) *Chamaephyta suffrutiscentia* (semishrubs), e.g. *Ephdera alata* and *Calligonum comosum.*

(d) Phanerphytes: Only some nanophanerophytes have the ability to form mounds. The lower branches of these plants are usually prostrate, and consequently able to trap moving sand and other materials close to the surface of the ground. *Lycium shawii, L. europaeum, Nitraria retusa* and *Tamarix* spp. afford examples. The most common example in the Sahara is *Zizyphus lotus,* which forms what are locally known as *nebkas.* These are rounded dune structures that can reach 10 m in diameter and 3 m in height. The Arabic name *sidra* is used both for the plant itself and for the sand mass in and on which the

> thorn bush grows. The formation of *nebkas* in great number in cornfields can be considered to be an important indicator of an advanced state of desertification.

The formation of mounds has many ecological consequences (Batanouny and Batanouny 1968, 1969). These include modification of the rooting habit of the plants, affecting the soil properties, and plant–soil–water relationships, microclimatic conditions and the vegetation.

7.2.8
Production of Ephemeral Roots (Shoot-Borne Roots)

Reference should be made in this respect to the so-called rain roots. These are fine rootlets produced in response to light rain showers by the woody parts of the root, just below the ground surface. They may also be produced during periods of dew formation. These roots are ephemeral and eventually wither, reappearing shortly after a new supply of surface water has become available. Field observations on many perennials of the Sahara confirm the conclusion of Kausch (1965), who experimented with cacti and showed that, "When roots of cacti which have been kept dry for many months are re-moistened, absorbing roots are formed after a few hours. The transpiration rate increases rapidly within a day after re-moistening. This is an indication that as soon as the roots appear water absorption begins immediately, and rapid improvement of the water relationships in the plant occurs."

Even the old parts of the roots, which are lignified and covered with cork layers, rapidly form young rootlets whenever the surrounding soil is wet. It is therefore the total root system that takes up water, not only its younger parts. The root hairs, which are the most active water-absorbing elements, cover the young roots from base to tip and, contrary to what botanical textbooks may say, are not restricted totally to a narrow zone above the growing points. They appear a few hours after the first rain.

7.2.9
Sheath Roots (Rhizosheaths)

Reference may also be made to the so-called sheath roots described by Volkens (1887), Price (1938), Killian and Lemée (1956), etc. Volkens (1887) studied the anatomy of the desert plants in Egypt on a physiological basis, and described rhizosheaths as being an adaptation to desert dunes. These are found in desert plants especially grasses. Roots seem to exude a mucous material that cements sand particles and forms a sleeve-like structure ensheathing the root. These sheaths seem to protect small roots against desiccation. *Stipagrostis ciliata, S. hirtigluma, S. obtusa, S. plumosa, S. raddiana, S. scoparia, Pennisetum divisum, Lasiurus scindicus, Ammophila arenaria, and Panicum turgidum* are among the many grasses having these sheath roots. The study of the probable occurrence of mycorrhiza and their role in the water economy of the desert grasses such as *Panicum turgidum, Lasiurus sindicus* and *Pennisetum divisdum* is an important, but not well-explored field.

7.2.10
Resistance to Exposure Due to Wind Erosion

Due to the continuous removal of surface deposits at many sites in the desert, the roots of already established plants are eventually exposed (Figs. 23, 24). Desert plants, especially psammophytes, have roots that tolerate exposure for a long time under the harsh desert conditions. They may produce vegetative growth when moistened. Examples of such plant species that grow in the deserts of the Middle East include *Moltkiopsis ciliata, Convolvulus lanatus, Polycarpaea repens, Echiochilon fruticosum, Scrophularia hypericifolia, Artemisia monosperma* and *Cornulaca monacantha*. Danin (1991) recorded cases in which the exposed roots of *C.*

Fig. 23. Exposure of the root system of *Calligonum comosum* (Arabic: *Arta*) growing in the Nafud, Saudi Arabia

Fig. 24. Exposure of the roots of desert plants growing in the northwestern desert of Egypt

monacantha are 1–2-m long. That the plant tolerates exposure of its roots may be explained by the internal structure of the root. Roots of *C. monacantha* (like many other lignified Chenopodiaceae) have internal active phloem and xylem elements (Fahn et al. 1986), enabling transport to take place even when the roots are exposed (Danin 1991). Danin (1991) suggested that protection of the xylem and phloem from external desiccation enables these drought-resistant species to withstand the principal stress of this habitat.

7.2.11
Absorption from a Relatively Dry Soil

Xerophytes have higher osmotic pressure values than mesophytes. The high osmotic pressure of the root and shoot tissues may be considered as a special physiological asset that adds to the efficiency of absorption (Killian and Faurel 1936). Osmotic values may serve as an exact indicator of the water balance and, therefore, of the total water economy of the plant (Walter 1974). The osmotic pressure values are closely related to the water supply and decrease with increasing water supply (Batanouny 1980).

Fitting (1911) carried out experimental ecological research in the northwestern part of the Sahara on plants from Biskra, and was the first to show the importance of the osmotic pressure of the sap of desert plants. He studied plants of different groups and determined their osmotic values by the plasmolytic method using potassium nitrate and sodium chloride solutions. Killian and Faurel (1933, 1936) further investigated some of these plants at Beni Unif; the latter author used the cryoscopic method. Walter (1964) compared the results obtained by Fitting with those of Killian et al., whose data showed that the osmotic values of plants growing in saline habitats are high, e.g. *Atriplex halimus* (35.5–70 atm), *Limonium delicatulum* (20.7–48.6 atm), *Limoniastrum guyonianum* (49.6–53.4 atm) and *Frankenia pulverulenta* (40.6 atm). Perennial xerophytes have osmotic values lower than halophytes, e.g. *Peganum harmala* (14.9–34.4 atm), *Capparis spinosa* (18.4–21.3 atm) and *Fagonia glutinosa* (9.7–17.3 atm).

The relatively high osmotic pressure of desert plants enables their roots to overcome the high total soil moisture stress (TSMS) that develops under xeric conditions and to withdraw more water than plants with low osmotic pressures. Slatyer (1960) found that xerophytes, such as *Acacia aneura*, can recover without injury from a diffusion pressure deficit (DPD) of about 130 atm. The tomato, in contrast, is killed when the DPD reached 45 atm. Zohary (1962) considered osmotic value to be one of the chief factors affecting drought resistance. The osmotic pressure values are closely related to water supply and decrease with increasing availability of water (Kreeb 1963; Abdel Rahman and Batanouny 1964b; Abdel Rahman et al. 1972).

7.2.12
Burial of Plants by Wind and Allocation of Resources

Burial by wind-deposited sand is an environmental stress commonly encountered by plants of coastal plain, lake shoreline, and desert sand dune communities. The tolerance of specific species and the general physiological responses of plants to varying degrees of burial have been extensively studied in coastal and lakeshore dune systems (Maun and Lapierre 1984; Harris and Davy 1987; Zhang and Maun 1992; Maun 1996; Brown 1997), but few comparable studies have been conducted in desert ecosystems. Extrapolating results from coastal dunes to desert dunes is difficult because of the additional stresses caused by toxic ions in the soil, drought, and the extreme temperatures in arid regions (Brown 1997).

In an environment with mobile sand, tolerance to partial burial seems to be a requisite for dominant plant species. *Sporobolus spicatus*, *S. pungens*, *Stipagrostis scoparia* are part of an ever-growing list of species known to be tolerant of moderate levels of burial (Batanouny and Batanouny 1968). It has been found that these species are among those that are especially tolerant of burial. In his study of three species of dune plants, Brown (1997) stated that all of them shifted their biomass above ground when buried. Similarly, after just 1 week of complete burial, the dune grass *Elymus farctus* responded by shifting a higher proportion of its biomass above-ground (Harris and Davy 1987).

Clearly, a buried plant must find the resources to extend its growing points above the sand surface, but the ability to maintain photosynthetic capability within the above-ground biomass may also be important to surviving burial.

Desert dune plants survive burial in much the same manner as coastal dune plants. Plasticity of resource allocation and the ability to overcome low net carbon assimilation rates are important for surviving burial. There is a great need for further studies that focus on the physiology and mechanisms of plant responses to burial.

7.2.13
Anatomical Features Enabling the Root to Tolerate Edaphic Aridity

The supply of soil moisture to plants is very meagre and variable as a consequence of severe climatic aridity. The surface layers of the desert soil are almost dry throughout the year, except for a few days during the wet season. Subsurface layers of deep soils may include layers with limited available water. However, in dry years, which are not uncommon, the soil at different depths may be dry and have no available water for plants. Figure 14 shows that the available moisture ranges from nil to a maximum of 5.9 %. The major part of the soil profile in this figure shows that there is <2 % available water at a depth of about 12 m. Dryness of the soil around plant roots is not infrequent. Consequently, one might expect that the root moisture could be affected. Studies dealing with the water content of the roots of desert plants are limited. However, the work of Fahn et al. (1986) has contributed to our knowledge in this field. As already mentioned, the presence of internal active phloem and xylem elements protects the roots from the harsh atmospheric conditions when they are exposed. Bahardwaj and Gopal (1979) stated that, even during the summer when the soil moisture content at various depths decreases considerably, root moisture is not as much affected. Oppenheimer (1960), Stocker (1960) and Migahid (1962) have described water storage tissues in the roots of various desert plants and other xerophytes. In *Haloxylon scoparium* (syn. *Haloxylon articulatum*; Killian 1941) the peripheral layer may become hard and corky, thus protecting the plant from the pressure of hard soil and from desiccation. The inner cortex may instead become spongy, forming dead sleeves that function as water storage tissue. Bahardwaj and Gopal (1979) reported that the roots of *Lygeum spartum* are strongly hygroscopic, absorbing 100 % of their original weight from moist air. They found that in *Tephrosia apollinea*, the cortex appears to be a water storage tissue and is protected by a thin periderm. They also noted that the width of cortex increases along with secondary growth in older plants. The authors postulated that water stored in the root cortex can be utilised for growth and reproduction during the dry period, when other climatic conditions are favourable

CHAPTER 8

Adaptations That Reduce Water Loss

As in any ecological system, the balance sheet of the water economy of desert plants has a debit and a credit side. The credit side is represented by the water sources available to the plant and its ability to absorb this water, while the debit side is the water lost by the plant. In order to survive under drastic edaphic and atmospheric droughts, desert plants must intensify water absorption, as already explained, and reduce water loss.

Before discussing the mechanisms of the adaptations of desert plants that reduce water loss, let us first consider the processes by which plants lose water. Transpiration is the main process through which water is lost. It occurs by way of the green parts, which have stomata (in the case of stomatal and cuticular transpiration), or from older parts over which cork has been laid down (in the case of lenticular transpiration). The latter type is, however, of limited value. From a physical point of view the rate of transpiration depends on the vapour-pressure gradient from the transpiring surface to the surrounding air. One can imagine how very steep this gradient must be under desert conditions of high vapour pressure deficit (VPD) and high temperature. All climatic conditions that affect evaporation influence transpiration across this vapour pressure gradient. Transpiration takes place rather as one would expect it, especially during the dry season and when the plant has green organs, which may be stems, leaves, or both. The greater the surface of these green organs, the higher will be the water loss of the whole plant. Reduction of the transpiring surface is an effective means by which many desert perennials cut down their water loss and keep a positive water balance. Reduction of the rate of transpiration, both daily and seasonal, is another mechanism of adaptation that decreases water loss. In this case, the total water output from the whole plant is also less.

8.1 Reduction of the Transpiring Surface

The most effective single adaptation by which desert plants reduce their water loss and adjust to variations in the amount of available water involves changes in the size of the transpiring body. In other words, the plasticity of the responses of desert plants to the changing soil water conditions is achieved mainly by changing the area of the transpiring surface. When a plant is confronted with a severe decline in the available soil moisture, it reduces its transpiring surface. This response can be continued until, in some cases, only a few living buds remain enclosed in dry dead leaves. When the plant is supplied with water in the next wet season, it produces new growth (Fig. 25). The growth and death of various organs of the shoot system, accompanied by this type of shedding, correlate with the availability of moisture to the plant.

Seasonal reduction in the transpiring surface is an important factor in maintaining the water balance of certain xerophytes (Orshan and Zand 1962). Orshan (1963) stated that seasonal dimorphism is a common feature among the chamaephytes and is generally associated with a seasonal

Fig. 25. The new growth of *Zilla spinosa* (Arabic: *Silla*) after a rainfall has replaced the already shed, dry old growth

reduction in their transpiring surface. The degree of reduction is largely a function of the specific annual moisture conditions. For some chamaephytes it has been demonstrated that this reduction is decisive in balancing water economy (Orshan 1953; Zohary 1962). These plants exhibit great plasticity regarding both the extent of their annual growth and the reduction of the transpiring body area (Orshan and Zand 1962). Among the chamaephytes growing in deserts of the Middle East that exhibit seasonal dimorphism, one might mention *Zygophyllum dumosum*, *Z. coccineum*, *Artemisia herba-alba*, *A. monosperma*, *Thymus capitatus*, and *Convolvulus lanatus*. Orshan and Zand (1962) stated that seasonal dimorphism is actually a special kind of drought-evading mechanism through which an increasing part of the transpiring plant's shoot is gradually shed as water shortage becomes accentuated during the long and dry summer, although the plant as whole retains its normal activity. They proposed the term "partial drought evasion" for this phenomenon. It has been found that the values of the relative seasonal reduction, i.e. the transpiring parts shed during summer, expressed as a percentage of the plant's transpiring area in late spring are greater among desert species than in the Mediterranean ones. Some desert plants, e.g. the bean caper (*Zygophyllum dumosum*), reduce their transpiring surfaces by up to 96% (Orshan 1963).

Orshan (1953) tried to evaluate plant life forms with respect to drought resistance and classified them according to the organs shed seasonally and the parts that persist throughout the year and are thus exposed to drought during the dry season. He distinguished four types:

1. *Leaf shedders*: Plants that shed only their leaves and inflorescence shoots seasonally
2. *Branch shedders*: Plants that shed parts of their branches seasonally
3. *Whole-shoot shedders*: Plants that shed the whole of their above-ground parts each season
4. *Annuals*: Plants that die each year

One or more of the adaptive mechanisms discussed below can achieve reduction of the transpiring body.

8.1.1 Shedding Part of the Green Cortex

The cortex is the transpiring and photosynthesising tissue and is shed by a number of species, e.g. *Haloxylon salicornicum* (Fig. 26), *Anabasis setifera* (Fig. 27), *Arthrocnemum macrostachyum* (Fig. 70), *Z. coccineum*, *Zygophyllum qatarense* (Figs. 28–30), and many succulent xerophytes and halophytes. In these species, the core of the young branches during the rainy

Fig. 26. *Haloxylon salicornicum* (Arabic: *Rimth*) branches showing the decortication of their old parts. This species is widespread in the deserts of the Middle East

Fig. 27. Decorticated branches of *Anabasis setifera* growing in the coastal plains along the Arabian Gulf

Fig. 28. A branch of *Zygophyllum qatarense* (Arabic: *Harm*) showing the shedding of the cortex from the old parts of the branch

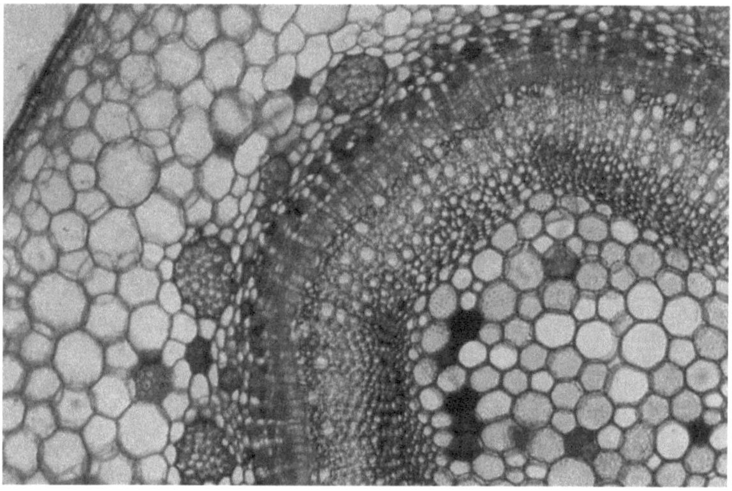

Fig. 29. Transverse section of a stem of *Zygophyllum qatarense* showing the succulent cortex

season consists only of a central cylinder containing the vascular bundles and surrounded by the cortex. The latter is composed of many succulent layers of large watery cells and photosynthetic layers of green cells and is covered by an epidermis having stomata (Fig. 29). As long as water is available, these cortical tissues remain green and perform the functions of photosynthesis and transpiration. When soil water becomes scarce but water continues to be lost through transpiration, cork is formed, separating the cortex cells from the main plant body (Fig. 25). Since cork is impermeable to water, the whole cortex is cut off from its water supply, dries out and is shed, and only the youngest internodes of the branches retain their photosynthesising cortex (Fig. 28). Without their green cortex the internodes cannot photosynthesise, but since they are also covered by cork they do not lose water by transpiration either. Only the few younger internodes that retain their green cortex continue to perform both functions. The seasonal reduction in the transpiring surface and the consequent decrease of water output brought about by cortex shedding is very considerable. It enables plants to survive the summer drought in an active state without becoming completely desiccated. Chenopodiaceae, which shed their cortex as well as some of their branches, produces flowers in autumn, at the end of the dry season. When they are artificially watered during the summer or grow in habitats where water is available even during the dry season, cortex shedding

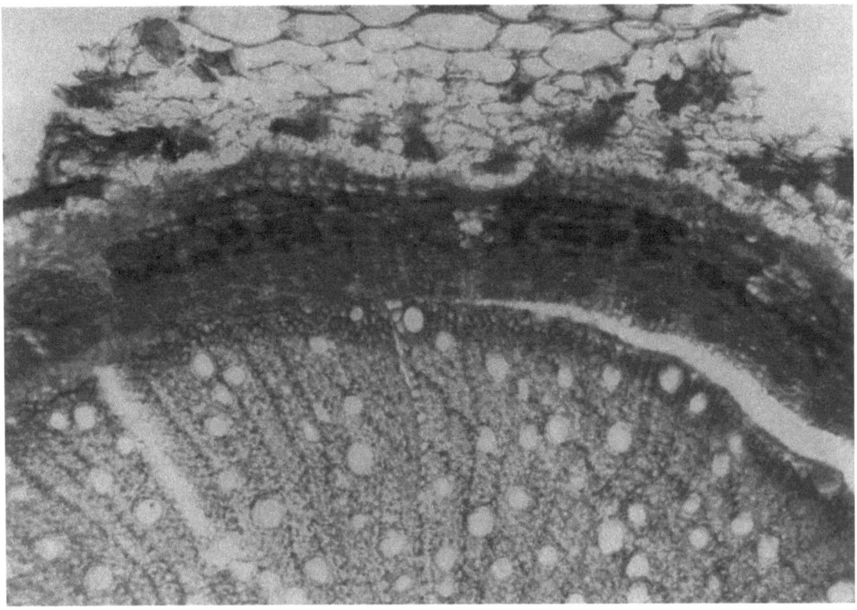

Fig. 30. Transverse section in the old decorticated branch of *Zygophyllum qatarense*

is much retarded and may even not occur at all. Evenari et al. (1971) reported that the degree of seasonal reduction of the transpiring surface is largely a function of specific annual moisture conditions. In drought years it is much more pronounced than in those years with an adequate water supply.

8.1.2
Shedding, Partly or Wholly, of Branches

This occurs in many members of the goose-foot family Chenopodiaceae and in other Families that are also cortex shedders. When the water available to the plants becomes very scarce, the whole plant does not die, rather only some of its branches. The branches that remain alive continue to carry out photosynthesis and transpiration through the activity of their various green tissues. With progressive drying of the soil, more and more branches die, leaving only a few or sometimes just living buds protected by dry leaves. After the rains in the subsequent wet season, the plant regenerates itself from these dormant renewal buds. This is evident in the case of *Cymbopogon shoenanthus* (Fig. 31). This illustrates how partial death, for the sake of survival, is an important adaptation in the life of desert plants (Evenari et al. 1971; Migahid et al. 1972).

Fig. 31. New growth of *Cymbopogon shoenanthus* (Arabic: *Idhkhir*) after rainfall. Note the dry parts from the previous years. Saudi Arabia, Mecca–Medina road

8.1.3
Leaf Shedding

Leaf shedding is an effective device which xerophytes that remain active during the dry season have developed to balance their water economy. Some desert plants are leafless almost all the year-round (e.g. *Retama raetam* and *Leptadenia pyrtechnica*, Fig. 32; *Calligonum comosum*, Fig. 33). Many other desert plants are leafless in the dry season (e.g. *Zilla spinosa*, Fig 34). Leaf shedding may be accompanied by branch shedding in some species. In such cases, the weight of the plant may be considerably reduced. The white broom *(Retama raetam)* provides an example of leaf shedding. This plant produces tender green shoots carrying a few small leaves that are soon shed under moisture stress, while the shoots remain as the only photosynthetic organ of the plant. All other broom-like xerophytes, such as *Leptadenia pyrotechnica*, behave in the same manner. When leaf shedding is combined with branch shedding, as in the thorny saltwort (*Noaea mucronata*), a 90 % or greater loss (Evenari et al. 1971) may vastly reduce the weight of the plant. *Zilla spinosa*, a desert undershrub growing in many countries of the Middle East, provides a good example of the effect of water supply on the persistence or shedding of leaves (Fig. 34). As a seedling, the plant has broad leaves, which wither over time. When conditions are still favourable in the wet season, the new growth has broad leaves and soft spines. At the advent of the dry season, however,

Fig. 32. A leafless branch of *Leptadenia pyrotechnica* (Arabic: *Markh*). United Arab Emirates

Fig. 33. A leafless shrub of *Calligonum comosum*. (Arabic: *Arta*)

Fig. 34. A leafless branch of *Zilla spinosa*. The plant is widespread in the deserts of the Middle East

leaves are shed, while the spines are lignified and become stiff. The whole plant becomes spiny and without leaves. In dense shrubs, the basal part shaded by the rest of the plant keeps its green leaves. Plants growing in shaded habitats can carry many more leaves in the dry season than those growing in the open.

8.1.4 Shedding of Winter Leaves and Retention of Summer Leaves

This is a device for reducing the transpiring surface in the dry season, e.g. in the gray sagebrush *Artemisia herba-alba*, *A. monosperma* and *Convolvulus lanatus*. Some plants only shed their large winter leaves and retain their small summer ones. For example, at the beginning of the rainy season *A. herba-alba* carries large pinnatifid winter leaves. These are shed at the beginning of the dry season and the new leaves formed become increasingly smaller and less pinnatifid as the season advances. The last leaves to be developed are merely tiny scales, which in years of severe drought are the only active parts of the plant. *A. monosperma* has the ability to produce minute simple leaves, 1–2 mm long, in summer and under drought conditions, as well as much longer and wider leaves, up to 30 mm long, under moister conditions (Danin 1996). The body reduction thus effected is very considerable. Winter and summer leaves differ not only in size and shape but also in their anatomical structure. The bean caper (*Zygophyllum dumosum*) has developed an especially effective method to reduce its transpiring surface by up to 96%. With increasing summer drought it sheds its leaf blades, leaving only the petioles. In another species, *Z. coccineum*, the leaflets are shed leaving the petioles, which can also be shed when conditions become drier.

Water loss through leaves is reduced in a wide variety of different ways including:

1. Reduced leaf area; the plasticity exhibited by desert plants regarding leaf area is astonishing. A desert shrub like *Lycium shawii* is a good example (Fig. 35). In wet habitats, it produces broad, thin leaves, while in dry sites it produces small leaves. Aggravation of drought results in the complete disappearance of leaves. Even the spines, which are weak and soft under conditions of ample water supply, become stiff and woody under dry conditions. It is noteworthy that for more than 1500 years, the Arabs have been able to distinguish between the two ecotypes of *Lycium* and *Ziziphus* that grow in dry and wet habitats.
2. Abscission of the leaves; complete leaf shedding has been discussed above. In some grasses, the leaves become desiccated from their tips in a basipetal direction when the plant is subjected to severe drought. This

Fig. 35. Two branches of *Lycium shawii* (Arabic: *Awasaj*) (from Qatar) growing in different habitats in the same locality and showing widely different leaf morphologies, especially regarding area. To the *right*, a branch of *Lycium shawii* growing on the dry slopes of rocky ridges. Note the very minute leaves

reduces the transpiring surface. Many grasses are also capable of persisting throughout the dry season, or even for several rainless years, with just green stems covered by green sheaths and dry laminas (Danin 1996). If the period of drought is extended, the leaf sheath may also die. *Cymbopogon schoenanthus* may dry out almost completely except for a few buds protected by dead organs and situated near the ground surface. After it rains, the plant starts to grow from these perennial buds and green circles of vegetative growth appear (Fig. 31).

3. Possession of a thick, waxy epidermis; this is a common feature of desert plants, as is the development of a multiple epidermis. Though the thickness of the cuticle is important, its chemical composition and consequently its role in minimising water loss are even more so.
4. Positioning of the stomata such that they avoid the direct impact of the climate; this can be achieved by rolling the leaf up under conditions of water stress. In this way, the abaxial side of the leaf, which is protected by a thick cuticle and is devoid of stomata, faces the harsh conditions of the

desert atmosphere. The stomata are on the inner surface of the tightly in-rolled leaf. This phenomenon can be observed in many desert grasses, e.g. *Ammophila arenaria, Lasiurus sindicus* (syn. *L. hirsutus*), *Pennisetum divisum, Stipagrostis* spp., *Halopyrum mucronatum,* and *Sporobolus arabicus* (Fig. 36). Rolling is usually accomplished by contraction of the bulbiform cells (Fahn 1989), which are found only on the adaxial surface and develop in lines parallel to the long axis of the leaf (Fig. 30). After shedding of the leaflets, the remaining petioles of *Zygophyllum dumosum* undergo a definite structural change. When carrying their blades, the petioles possess a thin-walled, one-layered epidermis. The stomata are large and lie in the epidermal plane After the leaf blades have been shed, however, the epidermal cells elongate and divide one or more times, their outer cell walls thicken, and the stomata are pushed below the epidermal surface. In this state the movement of the stomatal cells is extremely restricted – full opening of the aperture is impossible and transpiration is considerably reduced.

5. Orienting the leaves vertically to minimise the direct radiation they receive. *Erodium glaucophyllym* is a leafy perennial herb with broad leaves that are fully exposed to the sun's rays in the morning. Before noon, when the evaporative power of the atmosphere becomes high, the leaves

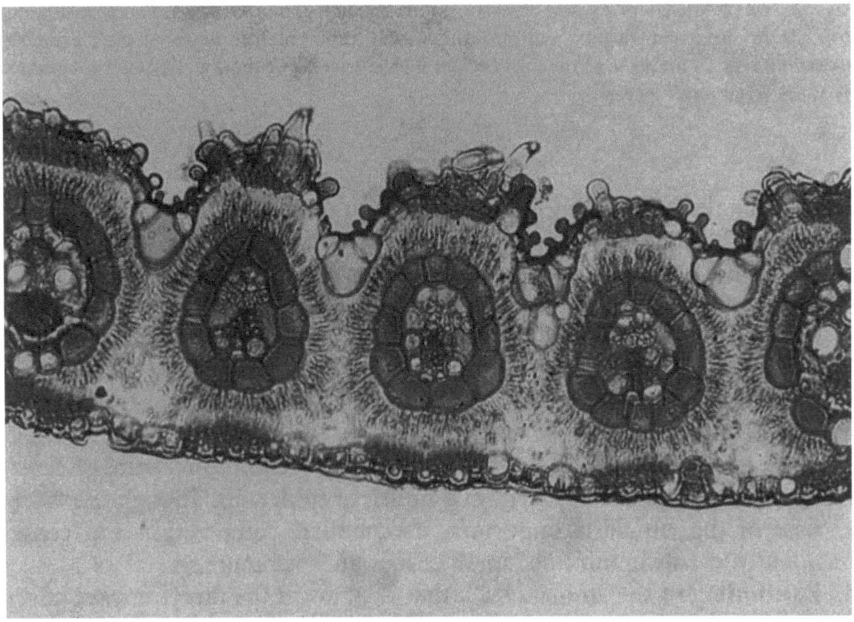

Fig. 36. Transverse section from the leaf of *Sporobolus arabicus* showing the hinge cells and the sheath of chlorenchyma surrounding the vascular bundle

are oriented so that their blades are parallel to the sun's rays. The same phenomenon is to be observed in the different *Senna* spp. growing in the deserts of the Middle East.

6. Investing the leaf surface with a covering of hairs (Fig. 37). Hairs have been credited with performing various functions; they can be regarded as an adaptation to strong solar radiation (Fahn and Cutler 1992; Barbour et al. 1993) and to sea spray in coastal regions (Boyce 1954) they also protect dune plants from the abrasive impact of airborne sand grains (Danin 1996). Many desert plants have a thick covering of hairs, e.g. *Arnebia hispidissima, Echiochilon fruticosum, Anchusa* spp., *Moltkiopsis ciliata,* other species of Boraginaceae, and *Aerva javanica* (syn. *A. tomentosa*) (Figs. 38, 39). Figure 39 shows the dense hairs on the surface of the stem of *Aerva javanica* (syn. *A. tomentosa*). *Aerva* (Arabic name *Ra'a)* is a species that occurs widely in the desert wadis of the Middle East. The Bedouin used to collect the coriaceous branches for stuffing cushions. Nowadays, it is rare to find such traditional cushions at the market. However, some Bedouin still keep them as part of their Arabic heritage.

8.2 Reduction of Transpiration

The transpiration of many desert plants under natural conditions has been investigated by numerous workers in the Middle East (Stocker 1928a; Seybold 1930; Evenari and Richter 1937; Evenari 1953; Migahid and Abdel Rahman 1953; Hammouda 1954; Migahid 1962; Abdel Rahman and Batanouny 1964a, 1965a,b; Migahid et al. 1972; Lange et al. 1969; and others). Although the average transpiration rates of desert plants are generally lower than those in mesophytes growing under favourable conditions, one cannot infer that desert plants are invariably characterised by low rates of transpiration. In many cases, when calculated on the basis of surface area, desert plants can be shown to have considerable transpiration rates. Irrigation of desert plants results in a considerable increase in their transpiration rate (Abdel Rahman and Batanouny 1965b). However, the transpiration rate per se is not very significant. The time over which high rates are maintained is more important, either over the day or over the year. In other words, most important is the ability of a plant to control its transpiration rate and reduce water loss, especially when there is a water deficit within both it and the soil. Perennial plants of the desert are able to control water loss by one or more of the devices mentioned, or by reducing their transpiration rate, either on a daily or a seasonal basis.

Fig. 37. A plate showing the hairs and trichomes covering the surface of different desert plant species as drawn by Volkens in 1884 (after Volkens 1887). *1, Lycium shawii; 2, Cocculus pendulus; 3, Paronychia arabica subsp. arabica; 4, Herniaria hemistemon; 5, Nauplius gtraveolens (Asteriscus graveolens); 6, Roemeria hybrida subsp. decandra; 7, Leontodon hispidulus; 8, Picris coronopifolia; 9, Monsonia nivea; 10, Stachys aegyptiaca; 11, Farsetia aegyptia; 12, Lotus halophilus; 13, Astragalus trigonus; 14, Phagnalon barbeyanum; 15, Ifloga spicata; 16, Pulicaria undulata; 17, Echinops spinosissimus; 18, Atractylis flava; 19, Convolvulus lanatus; 20–21, Plantago cylindrica; 22–23, Bassia muricata; 24, Cotula cinerea; 25, Anthemis melampodina; 26, Launaea nudicaulis; 28, Silene linearis; 29, Lavandula pubescens; 30, Lappula spinocarpa; 31, Achillea fragrantissima; 32, Centaurea aegyptiaca; 33, Ononis serrata; 34, Fagonia arabica; 35, Pulicaria crispa*

Fig. 38. A branch of *Aerva javanica* (Arabic: *Ra'a*) covered by the thick hairs

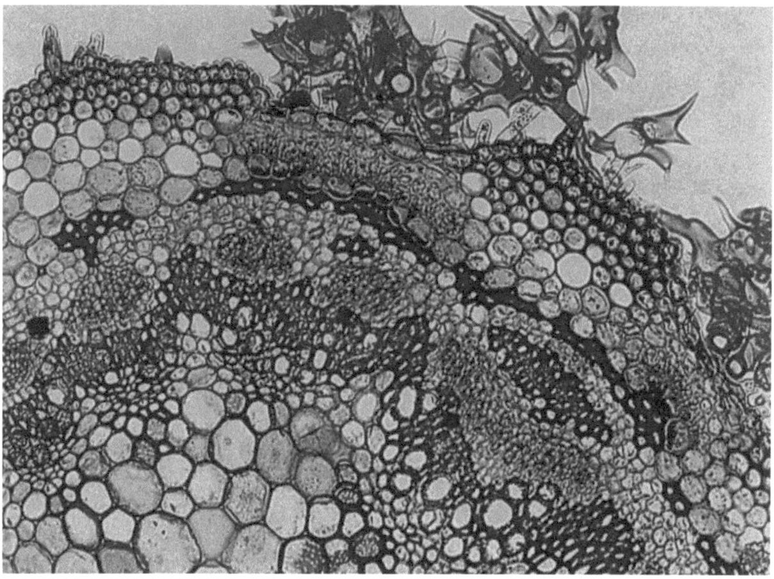

Fig. 39. A transverse section from the stem of *Aerva javanica* showing the dense hair cover on the surface

8.2.1 Daily Reduction of the Transpiration Rate

The daily fluctuations in the transpiration rate of numerous perennial desert plants has been extensively investigated, e.g. in *Retama raetam* (Hammouda 1954), *Zygophyllum coccineum*, *Zilla spinosa*, and *Pennnisetum divisum* (syn.*P. dichotomum*) (Abdel Rahman and Batanouny 1964a), and in *Leptadenia pyrotechnica* (Migahid et al. 1972), *Stachys aegyptiaca* (Batanouny 1981a), and many other species. We shall now consider five examples of desert perennials that represent various ecological types: (1) leafless shrubs, e.g. *Leptadenia pyrotechnica* (Fig. 40); (2) succulent shrubs, e.g. *Zygo-*

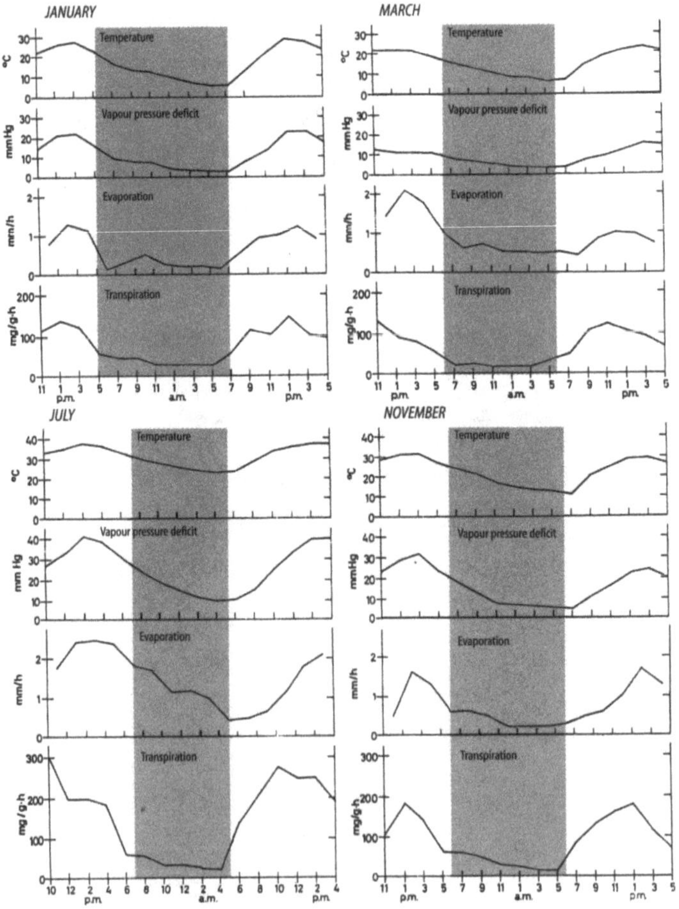

Fig. 40. Daily changes in the transpiration rate of *Leptadenia pyrotechnica* (mg/g fresh weight per h) and the climatic factors in the different months. The *shaded area* represents the night-time

phyllum coccineum (Fig. 41); (3) leaf-shedding spiny undershrubs, e.g. *Zilla spinosa* (Fig. 42); (4) grasses, e.g. *Pennisetum divium* (syn. *P. dichotomum*) (Fig. 43); and (5) leafy undershrubs, e.g. *Stachys aegyptiaca* (Fig. 44). The data presented in these figures demonstrate the following:

1. In spite of the severe environmental conditions under which these five types of plants live, their transpiration rates are low, with maximum values in summer reaching 300 mg g^{-1} h^{-1} in *Leptadenia* (Fig. 40), 161 mg g^{-1} h^{-1} in *Zygophyllum* (Fig. 41), 471 mg g^{-1} h^{-1} in *Zilla spinosa* (Fig. 42), 435 mg g^{-1} h^{-1} in *Pennisetum* (Fig. 43), and 720 mg g^{-1} h^{-1} in

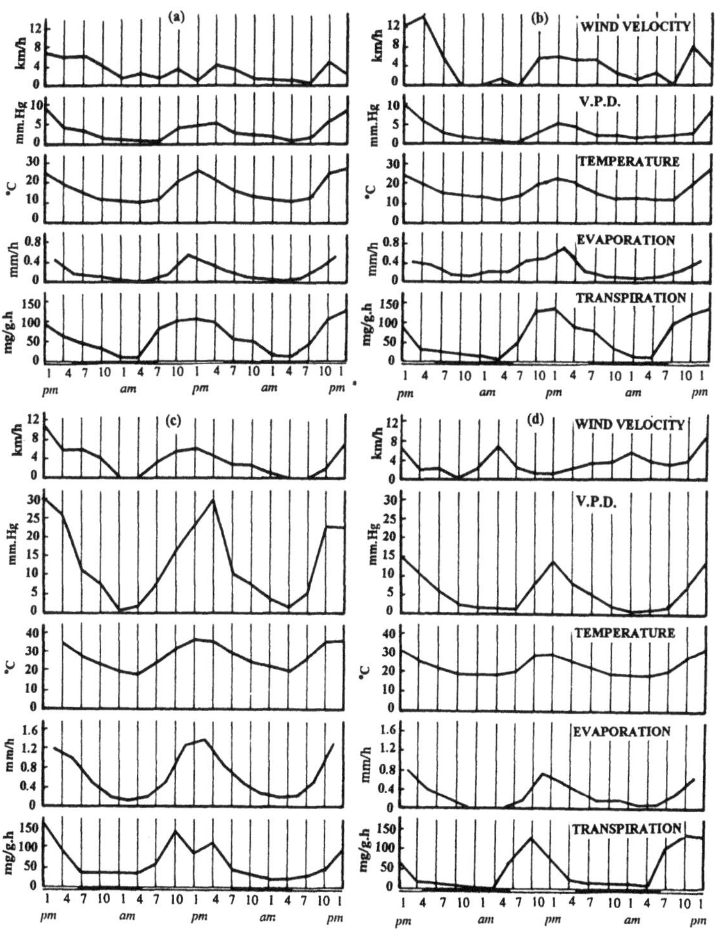

Fig. 41a–d. Diurnal fluctuations in the transpiration rate of *Zygophyllum coccineum* and the climatic factors in the different seasons. Wadi Hoff, the Egypto-Arabian Desert, 1960. **a** February; **b** Mach; **c** July; **d** November. *VPD* Vapour pressure deficit. (After Batanouny 1963)

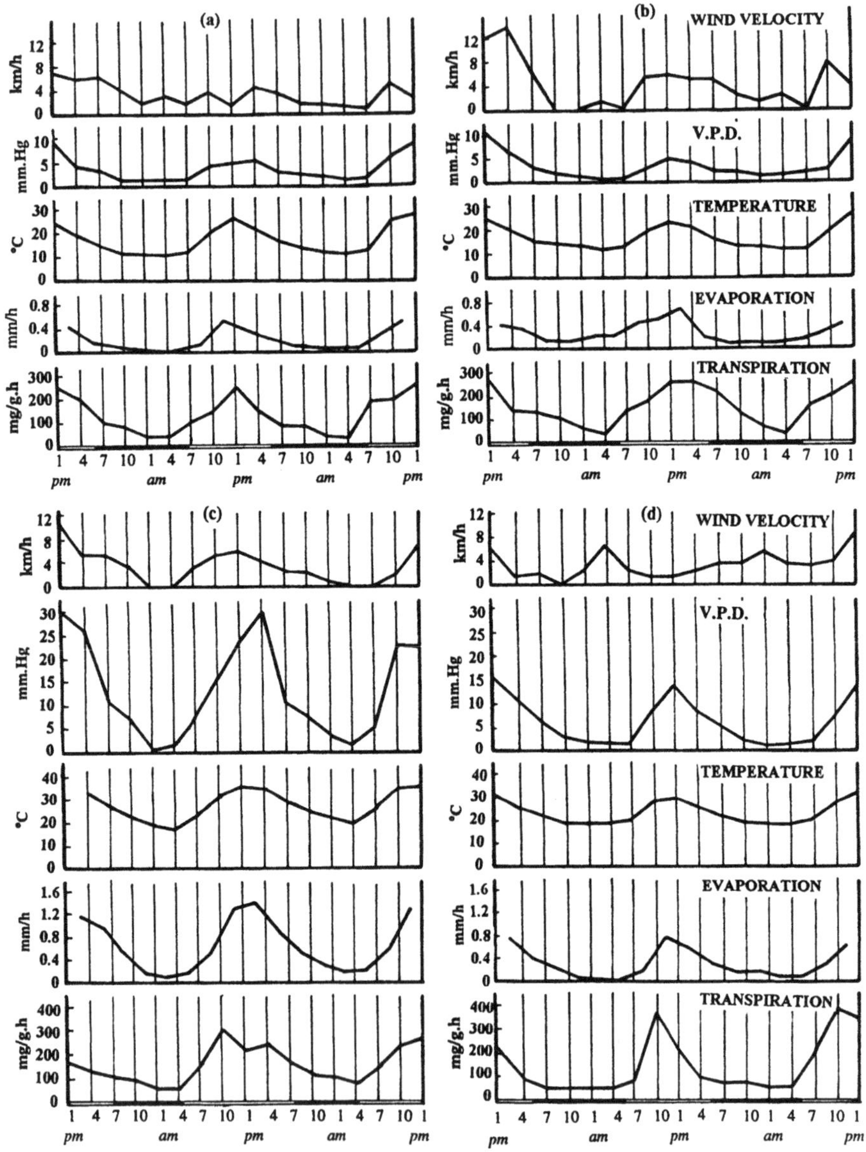

Fig. 42. Diurnal fluctuations in the transpiration rate of *Zilla spinosa* and the climatic factors in the different seasons. Wadi Hoff, the Egypto-Arabian Desert. **a** February; **b** March; **c** July; **d** November. *VPD* Vapour pressure deficit. (After Batanouny 1963)

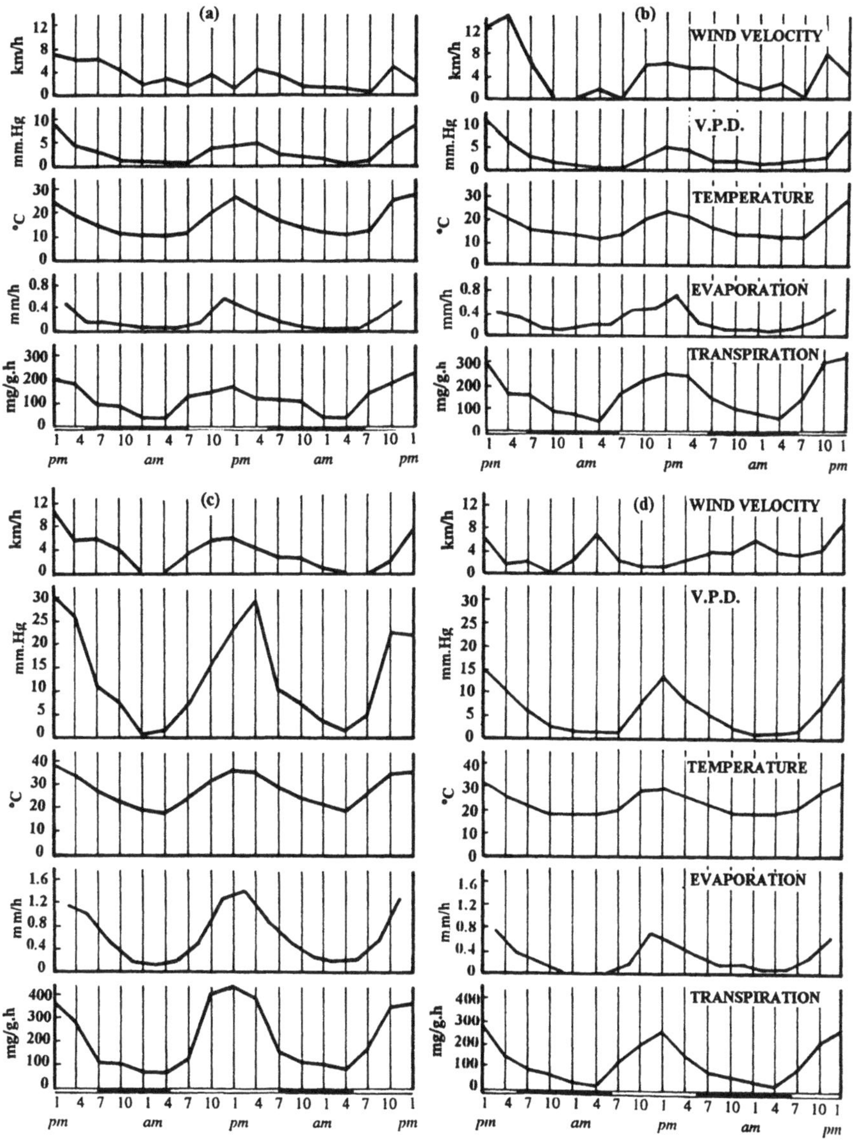

Fig. 43. Diurnal fluctuations in the transpiration rate of *Pennisetum divisum* (syn. *P. dichotomum*) and the climatic factors in the different seasons. Wadi Hoff, the Egypto-Arabian Desert, 1960. **a** February; **b** March; **c** July; **d** November. *VPD* Vapour pressure deficit. (After Batanouny 1963)

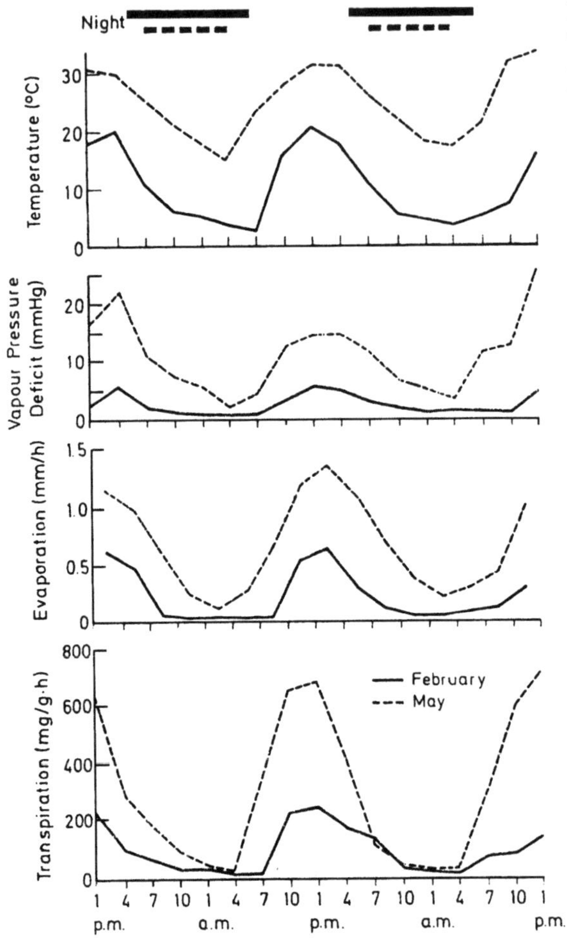

Fig. 44. Daily changes in the transpiration rate of *Stachys aegyptiaca* Pers. (mg/g fresh wt per h) and the climatic factors in February and May under natural conditions. (After Batanouny 1981b)

Stachys (Fig. 44). *Zygophyllum*, with its succulent leaflets, petioles and cortex, has the lowest transpiration rate. Leafless shrubs (*Leptadenia*) have lower transpiration rates than grasses, spiny shrubs and leafy undershrubs. It should be emphasised that these values are the maximum recorded. The duration of such high values during the daytime is limited.

2. In all species investigated, the transpiration rate is very low just before sunrise in every month, then rises sharply at sunrise. This occurs despite the fact that the evaporative power of the atmosphere is close to its minimum level at that time (Fig. 45). This clearly indicates that light-induced opening of the stomata causes the increase in transpiration in the early morning hours.
3. There is an evident lack of parallelism between the daily curves of transpiration and those of the external atmospheric factors. This is due to

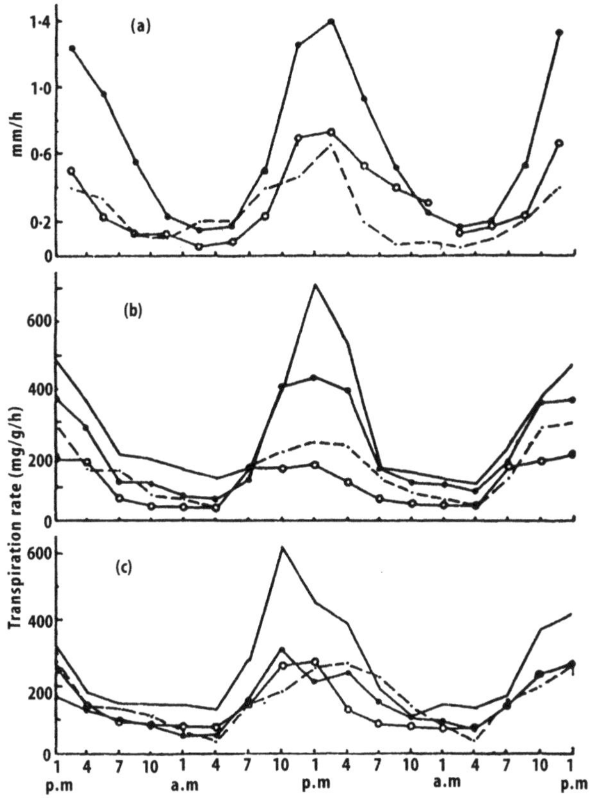

Fig. 45. Diurnal fluctuations in **a** the evaporation rate, and the transpiration rates of **b** *Pennisetum divisum* (syn. *P. dichotomum*) and **c** *Zilla spinosa* under different environmental conditions. *Filled circles* July 1960; *smooth line* watered plants, July 1960; *dotted line* March 1960; *open circles* December 1960. (After Abdel Rahman and Batanouny 1965b)

the influence of internal regulating mechanisms (probably mainly a stomatal response) that tend to reduce transpiration during the day while atmospheric evaporation intensifies. The discrepancy between the curves of transpiration rate and those of atmospheric factors is more pronounced in summer than in winter and spring . The maximum rate is usually attained before noon and is followed by a considerable decline during the afternoon hours.

4. The daily fluctuations in transpiration in desert plants is reflected in the different types of transpiration curves and, in turn, on prevailing environmental conditions. Batanouny (1974) distinguished the following types of transpiration curves in the case of *Zilla spinosa* (Fig. 46):
   a) A *dome-shaped curve* is obtained under mild climatic conditions and relatively considerable water supply. These conditions usually occur in the wet season. They show neither a sharp rise in the forenoon hours nor a pronounced downward trend during the rest of the day. The curve has no flat minimum at night. The discrepancy between the transpiration and evaporation curves is not pro-

nounced. Similar curves have been obtained for *Zygophyllum coccineum* (Abdel Rahman and Batanouny 1964a, 1965b).

b) In a *single-peaked curve*, the time of maximum transpiration depends on the soil moisture content, the water conditions in the plant body and the period of increased activity of the plant, as indicated by vegetative growth or flowering. Two subtypes have been recognised: (1) a curve with the peak of transpiration reached simultaneously with that of evaporation. This occurs when there are favourable soil moisture conditions and a high evaporation deficit in the atmosphere. It is usually observed during the late spring and early summer in rainy years. (2) A curve with maximum transpiration attained earlier in the day than maximum evaporation occurs when the soil moisture is depleted and climatic factors are still severe. Such conditions are met with at the end of the dry season. In this context, Oppenheimer (1951) wrote: "A remarkable feature of the daily curves of transpiration in desert plants is the frequency of curves with a steep ultramaximum in

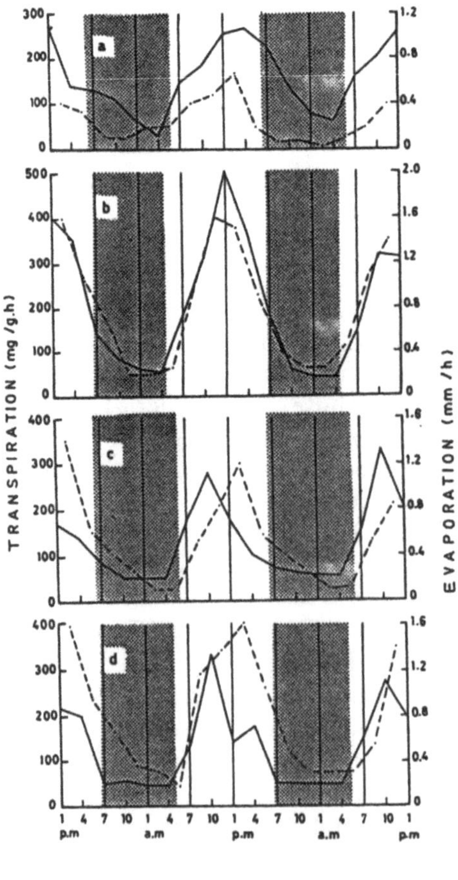

Fig. 46a–d. Daily changes in transpiration of *Zilla spinosa* (*smooth line*) and evaporation (*dashed line*) in different months. **a** Dome-shaped curve in March; **b** single-peaked curve in May; **c** single-peaked curve with early maximum in September; **d** double-peaked curve in August. The *shaded area* represents the night-time. (After Batanouny 1974)

the early morning and pronounced downward trend to low figures during the rest of the day."

c) *Double-peaked curves* are characterised by a tendency of transpiration rates to fall during the daytime. This drop is followed by a rise. The second fall occurs during the night, reaching low levels after dawn. In general, the second peak on the same day is not as high as the first one. The earlier reduction in transpiration due to water deficit can provide an opportunity for the plant to recover. This leads to the second rise in the transpiration rate, during the remainder of the day. The double-peaked curves are usually observed in the mid-summer and autumn months, when the climatic and edaphic aridity is high.

It is clear that perennial desert plants behave differently with respect to their capacity to reduce their transpiration rate, but they are all able to control it.

8.2.2
Seasonal Reduction in Transpiration Rate

In some plants, such as *Zygophyllum coccineum*, the mean transpiration rate shows little variation over the year. This is in contrast to the wide variation in the evaporating power of the atmosphere (Abdel Rahman and Batanouny 1964a). In the case of *Leptadenia pyrotechnica* (Fig. 34), the magnitude of the monthly variation in the mean transpiration rate is less than that of the evaporating power of the atmosphere (Migahid et al. 1972). It was found that the highest mean transpiration rate is nearly twice the lowest one, while the highest mean evaporation rate is about four times that of the lowest (Migahid et al. 1972; Fig. 34). The rise in the transpiration rate during the summer months is not comparable to the rise in the evaporating power of the atmosphere. This is clearly demonstrated by *Zilla spinosa*, in which the highest mean transpiration rate is twice the lowest one, while the highest mean temperature, evaporation rate and vapour pressure deficit (VPD) values are respectively about 3, 11 and 22 times those of the lowest (Abdel Rahman and Batanouny 1964a). Such very wide differences in the relationship between transpiration rate and climatic factors point to the fact that *Z. spinosa* and other species like it exert strong control over water loss, maintaining the rate at a level that is not equivalent to that reached by atmospheric factors in the summer. Numerous investigators have reported a limitation of transpiration during both the dry period and under aggravated conditions of moisture stress.

Some desert plants show wide ranges of variation in transpiration rate, both daily and seasonal. Such plants have been referred to by Walter (1937) as "euhydric" plants, (e.g. *Pennisetum divisum*). At the same time, according to Walter (1937), other species are hydrostable or "stenohydric". They show a narrow range of variation in transpiration rates (e.g. *Leptadenia pyrotechnica).*

8.2.3
Reduction in Water Output of Individual Plants

After considering the reduction of both the transpiring surface and the transpiration rate in the dry season as adaptations that reduce water loss, it seems proper to consider the amount of water lost by individual plants in different seasons. Ervenari et al. (1971) stated that the real importance of these factors for the survival of desert plants could best be appreciated if the water loss from an entire plant were calculated for each month of the year. They stressed the enormous efficacy of the combined action of reduction in transpiring surface, seasonal decrease of transpiration rate, and morphological-anatomical changes in the transpiring organs in reducing water loss by the whole plant. Some of the data given by Evenari et al. (1971) are shown in Table 3. These data show that the two morphologically different species *Retama raetam*, a leafless desert shrub, and *Heliotropium rotundifolium*, with relatively large leaves, decrease their water output during the dry season to 1.8 and 3.5 % of their maximal values, figures that clearly indicate the outstanding adaptability of these summer-active xerophytes to the water conditions of their habitats.

Table 3. Monthly water output from entire individual plants of *Retama reatam* and *Heliotropium rotundifolium*. (After Evenari et al. 1971)

Month	Rainfall (mm)	*Retama raetam* Average output (kg/plant)	Maximum output (%)	*Heliotropium rotundifolium* Average output (kg/plant)	Maximum output (%)
January	7	180	47	1.2	16
February	74.5	350	92	5.6	75
March	4	380	100	6.1	81
April	9	360	95	7.5	100
May	0	108	28	1.8	24
June	0	23	6.1	0.55	7.5
July	0	9	2.4	0.6	8
August	0	10	2.6	0.4	5.3
September	0	11	2.9	0.34	4.5
October	0	7	1.8	0.23	3.5
November	4.5	21	5.5	0.7	9.3
December	8.5	130	34	0.9	12

8.2.4 Reduction of Water Output from Natural Vegetation

The total water output of desert vegetation was calculated by Abdel Rahman and Batanouny (1965a) in the different microhabitats of wadi Hoff (Fig. 9). On the plateau, the area studied was a waterway about 25 m above the wadi bed, covered with fine deposits and supporting an open plant cover. It represents a distinct microhabitat of barren eroded plateau. The wadi bed, lying between a vertical cliff on the western side and a slope on the eastern side, includes several other microhabitats, namely the shaded microhabitat, the wadi bed proper, and the first and second terraces.

The shaded microhabitat occupies the area made by the undercutting of the cliff and is characterised by accumulations of fine deposits on the soil surface and of plant cover protected by the protruding edge of the cliff. This microhabitat faces east and is shaded for most of the day.

The wadi bed proper is devoid of vegetation and acts as the main channel of the wadi. The soil is covered at the surface by small pebbles and coarse sand.

On the eastern side of the wadi bed there is a belt of fine sand and silt accumulations, 2–4 m in width. This belt forms an elevated terrace, about 50 cm higher than the wadi bed proper, and is termed the first terrace.

The first terrace is followed eastward by another, more elevated one, and a second terrace, which is rugged, with blocks and boulders on its surface. Fine soil lodges among these blocks and boulders. The main source of water in the Egyptian deserts, particularly in the area studied, is from rainfall. The average annual rainfall in the region studied is 31 mm. Runoff plays a very important role in plant life and its distribution. The unevenness of the land surface in the desert leads to irregular distribution of rainfall.

The total water output from a sample of desert vegetation depends on different factors, mainly the fresh weight of the plants, their density, the climatic factors, the floristic composition, and the availability of soil moisture; these factors may interact with each other. The diagrammatic representations of the cover and the water output (kg/day) of the plants in selected areas in the different microhabitats of the wadi, together with calculations of total fresh weight and water output per 100 m^2 in March and July, are given in Figs. 47 and 48. The water output in March, representing the wet season, ranged from 8.265 kg per 100 m^2/day in the second terrace with sparse vegetation to 114.240 kg per 100 m^2/day in the first terrace of the wadi with dense vegetation. Water output in the dry season in the two microhabitats ranged from 7.976 kg per 100 m^2/day in the second terrace to 119.990 kg per 100 m^2/day in the first terrace. This shows that there is a slight rise in water output in the first terrace in summer. However, such a rise is not comparable to the rise in the evaporative power of the atmosphere. The fresh weight of the plants in 100 m^2 is 46.825 kg in March and 44.300 kg in July. It seems

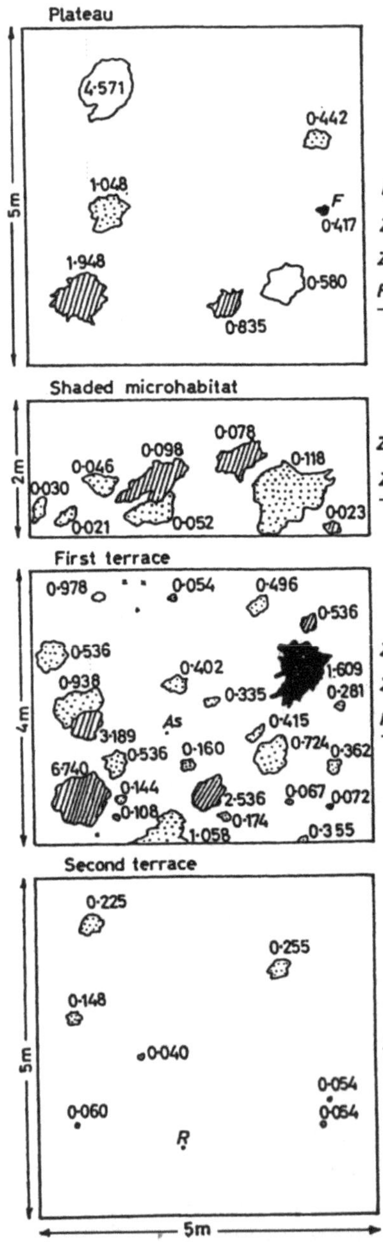

March 1960

Plateau		Fresh wt (kg/100m²)	Water output (kg/100m².day)
Iphiona mucronata	○	7·120	27·168
Zilla spinosa		3·200	11·132
Zygophyllum coccineum		4·320	4·808
Farsetia aegyptia	F ●	0·520	1·668
Total		15·160	44·776

Shaded microhabitat		Fresh wt (kg/100m²)	Water output (kg/100m².day)
Zygophyllum coccineum		37·200	26·800
Zilla spinosa		10·200	19·100
Total		47·400	45·900

First terrace		Fresh wt (kg/100m²)	Water output (kg/100m².day)
Zygophyllum coccineum		24·200	32·420
Zilla spinosa		19·625	71·095
Pennisetum dichotomum	P●	3·000	10·725
Total		46·825	114·240

Second terrace		Fresh wt (kg/100m²)	Water output (kg/100m².day)
Zygophyllum coccineum		6·150	8·265

Fig. 47. The plant cover and the water output (kg/day) of plants in selected areas in different microhabitats. The total fresh weight and water output per 100 m² in March 1960 are also given. (After Batanouny 1963)

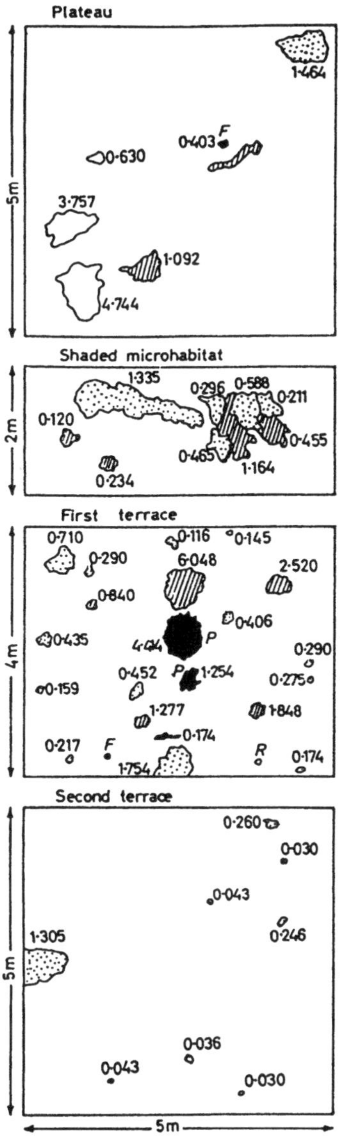

July 1960

Plateau	Fresh wt (kg/100m^2)	Water output (kg/100m^2.day)
Iphiona mucronata	7·720	34·120
Zilla spinosa	2·240	9·408
Zygophyllum coccineum	4·880	5·856
Farsetia aegyptia F●	0·480	1·612
Total	15·328	51·000

Shaded microhabitat	Fresh wt (kg/100m^2)	Water output (kg/100m^2.day)
Zygophyllum coccineum	33·900	28·950
Zilla spinosa	8·875	19·737
Total	2·775	48·687

First terrace	Fresh wt (kg/100m^2)	Water output (kg/100m^2.day)
Zygophyllum coccineum	20·000	28·985
Zilla spinosa	18·650	62·665
Pennisetum dichotomum P●	5·650	28·340
Total	44·300	119·990

Second terrace	Fresh wt (kg/100m^2)	Water output (kg/100m^2.day)
Zygophyllum coccineum	5·500	7·976

Fig. 48. The plant cover and the water output (kg/day) of plants in selected areas in different microhabitats. The total fresh weight and water output per 100 m^2 in July 1960 are also given. (After Batanouny 1963)

possible that the slight rise in the water output during the summer is due to the rise in transpiration rate, but it decreases to a level that is not comparable to the rise in the evaporation rate. On the second terrace, both the fresh weight and the water output are reduced. In this microhabitat there are only a few scattered individuals of *Z. coccineum*. They showed a decrease in fresh weight, from 6.150 kg per 100 m^2 in March to 5.500 kg per 100 m^2 in July. The ratio of water output per day to the total fresh weight of the plants in a definite area is widely variable. As is clear from Fig. 48, this ratio was about 3, 1, 2.4 and 1.3 in the plateau, shaded microhabitat, and first and second terraces, respectively. This is mainly attributable to differences in the climatic conditions and the floristic composition of the vegetation in the different microhabitats.

The total water output per day from an area of 100 m^2 is nearly equal on the plateau and in the shaded microhabitat despite great variation in the total fresh weight of plants in these two habitats. On the plateau the total fresh weight is only one-third of that in the shade. This difference is mainly attributable to the effect of shade in decreasing transpiration rates. Though the fresh weights are nearly equal in shade and on the first terrace, water output is 114.2 kg per 100 m^2/day on the first terrace and 54.9 kg per 100 m^2/day in the shade. The low water output can mainly be attributed to the effect of shade.

The lowest water output was recorded on the second terrace, where it amounted to 8.265 kg per 100 m^2/day. This may be due to both the sparse vegetation in this microhabitat and the dominance of *Z. coccineum* with its low transpiration rate. The fresh weights of the common plants in the different microhabitats are completely different (Figs. 47, 48). The difference between the transpiration rates of the various species and the wide variation in their fresh weights in the different microhabitats contributed to the great difference in water output. Though the fresh weight on the plateau amounts to 15.16 kg per 100 m^2, i.e. 2.5 times that on the second terrace (6.15 kg per 100 m^2), the water output of the former is more than 5.5 times as much as that of the latter. This can be explained by the presence on the plateau of species with higher mean transpiration rates than that of *Z. coccineum*. The species inhabiting the plateau and their transpiration rates are: *Iphiona mucronata* (158 mg g^{-1} h^{-1}), *Zilla spinosa* (150 mg g^{-1} h^{-1}), and *Farsetia aegyptia* (145 mg g^{-1} h^{-1}). *Z. coccineum* (49 mg g^{-1} h^{-1}) ranks next to *Iphiona* in density and total fresh weight. On the second terrace, *Z. coccineum* with its low transpiration rate, constitutes nearly the entire plant cover, resulting in a comparatively low water output.

In the shaded microhabitat, the water output per day (45.9 kg per 100 m^2) is less than the total fresh weight (47.40 kg per 100 m^2), in contrast to that found in the other microhabitats. This relatively low output may be attributed to two factors, namely, the high density of *Zygophyllum* with its low transpiration rate, and the effect of shade.

On the first terrace, the vegetation is denser than in the other microhabitats and the proportion of plants with high transpiration rates is greater. It is concluded that for these reasons the water output is higher than that in the other microhabitats. It amounts to 114.2 kg per 100 m^2/day.

Abdel Rahman and El Hadidy (1958) recorded a maximum water output in the dry period of 1955 of 16.0 kg per 100 m^2/day for a sample of desert vegetation along the Cairo–Suez road. This low figure is due to the sparseness of vegetation in the open gravel-covered desert, where the total fresh weight of perennials in February reached 3.21 kg per 100 m^2. This is very low compared to the weight recorded in the desert wadi.

8.2.5 Reduction of Water Output at the Community Level (Communal Reduction of Water Output)

It has been shown (Sect. 8.2.4) that the water output from natural vegetation varies according to a number of factors, e.g. floristic composition, the fresh weight of the plants and their density, the climate and the availability of soil moisture. The data shown in Figs. 47 and 48 reveal the effect of these factors in various habitats. However, one should also consider water output on the community level in the same habitat all year-round. The diversity of the various components of the desert plant community, as related to the transpiration rate, affect water output. This effect will be explained below.

8.2.5.1 Species Diversity

Transpiration rate and, consequently, water output varies in different plant species. Desert vegetation consists of a permanent framework of perennials, filled in during wet years by annuals. The density of these annuals depends on the amount and distribution of both rainfall and runoff water to the habitat. The plant cover during the rainy season can be considerable in wet years and is composed mainly of ephemeral plants, whose transpiration rate is high under such conditions (Abdel Rahman and Batanouny 1964a). In *Trigonella stellata*, transpiration ranged from a minimum of 270 mg g^{-1} h^{-1} at 04.00 h to a maximum of 1670 mg g^{-1} h^{-1} at 13.00 h, with a daily mean of 876 mg g^{-1} h^{-1}. In *Plantago ovata*, the transpiration ranged from a minimum of 183 mg g^{-1} h^{-1} at 04.00 h to a maximum of 820 mg g^{-1} h^{-1} at 13.00 h. In *Diplotaxis acris*, the minimum transpiration rate was 127 mg g^{-1} h^{-1} at 04.00 h, while the maximum was 1555 mg g^{-1} h^{-1} at 13.00 h. Perennial species do not show the same rates of transpiration. Abdel Rahman and Batanouny (1964a) recorded transpiration values for perennials in March that were much lower than those for annuals. In *Zilla spinosa*, they found that the transpiration rate ranged from a minimum of 32 mg g^{-1} h^{-1} at 04.00 h to a maximum of 273 mg g^{-1} h^{-1} at 13.00 h. In *Pennisetum divisum* (syn. *P. dichotomum*), the values ranged from a minimum of 36 mg g^{-1} h-1 to a maximum of 295 mg g^{-1} h^{-1} at the same times. The transpiration rate of succulent plants such as *Z. coccineum* is very low compared with that of non-succulent species. In *Zygophyllum*, the transpiration rate ranged from a minimum of 10 mg g^{-1} h^{-1} at 04.00 h to a maximum of 102 mg g^{-1} h^{-1} at 16.00 h.

Species diversity is reflected in water output, due to the fact that every species has its own level of water loss at a particular time of the year. High water output levels from some species coincide with low levels from others. This normally ensures a continuous supply of water throughout the year. The disappearance of some species or the diminution of their transpiring surface and transpiration rate reduces the water output. Thus, as the water supply decreases, some plant species disappear while others reduce their water output.

8.2.5.2
Diversity of Life Forms

Raunkiaer (1934) suggested that the location of a plant's renewal buds, as seen in various life forms, best expresses its adaptation to the season that is unfavourable for plant life. It has been claimed that ephemerism and persistence represent the two major strategies by which plants overcome the dry season (Evenari et al. 1975). Persistent plants, which maintain assimilating and transpiring organs during the entire summer, are "arido-active", while ephemeral plants are "arido-passive". Among persistent plants, chamaephytes (Ch) are considered to be better adapted to summer drought than are phanerophytes (Ph), since they partly reduce their transpiring and assimilating organs during the dry season (Orshan and Zand 1962; Orshan 1963). Ephemerals are more resistant to summer drought than the hemicryptophytes (H) and geophytes (G), because the former spend the summer as seeds and the latter as vegetative organs. Therefore, low values of the ratios Ph/Ch and (H+G)/Th might be assumed to express a higher degree of adaptation to drought. This applies to both chamaephytes among persistent species and to therophytes (see below) among ephemerals. In fact, both ratios increase with increasing precipitation (Danin and Orshan 1990).

If we consider the Egyptian deserts as typical of the deserts of the Middle East, it is significant that the study of Hassib (1951) described the life forms of no less than1800 species. (It should be noted that the number of species recorded today is 2071.) The main points of Hassib's study can be summarised as follows: The number of flowering plants in the inland desert is 775; these constitute 41 % of the total number of species representing the Egyptian flora. These plants are mainly therophytes (46.5 %). Chamaephytes (17.1 %) and hemicryptophytes (19 %) are also fairly well represented, but micro- and nano-phanerophytes include only 47 species (7.2 %). The other life forms are even less common: the geophytes represent 6.8 %, helophytes include 17 species (1.9 %) growing in wet places near wells, and parasites are represented only by seven species (0.9 %). Mega- and meso-phanerophytes are absent.

Schulze (1982) compared the carbon, water and nutrient turnover of annual and perennial herbaceous species, shrubs and trees, corresponding more or less to Raunkiaer's life forms. He concluded that it is not the unfavourable season, as suggested by Raunkiaer, but rather the carbon assimilation that develops during the favourable season that determine the success of a specific life form. Schulze suggested that adaptations to the environment occur through carbon allocation, which determines not only the structural form of plants, but also the seasonality and life span of their assimilating organs. Indeed, life forms are the outcome of growth processes, which are related to the morphogenetic processes of plants and are affected by carbon allocation during the growing season (Danin and Orshan 1990). However, the seasonal changes in plants, which are expressed by their life forms, play an important role in water, carbon, and mineral turnover (Evenari et al. 1975) and are not affected by carbon allocation only (Danin and Orshan 1990).

Within desert plant communities, it can be observed that the component species belong to various life forms according to the concept of Raunkiaer. Ephemerals, known as therophytes according to Raunkiaer, represent the largest number of species. However, it should be noted that the term therophytes does not apply to many species in the deserts of the Middle East. The majority of annuals are winter ones, with only a limited number of annuals found in the summer. These can truly be termed "therophytes", as their name implies. Finally, the presence of different life forms within a desert plant community indicates the occurrence of different patterns of water loss by transpiration. As ephemerals grow only in the wet season, they lose significant amounts of water during their short lives. However, albeit infrequently, they may not be present at all. The bulbous geophytes shed their transpiring surface (the aboveground leaves) before the advent of the dry season. Consequently, their water output ceases. Chamaephytes and hemicryptophytes posses various adaptations that minimise water loss throughout the year. This is particularly the case during the dry periods, in which the water supply decreases while the atmospheric conditions enhance water loss. Phanerophytes are not well represented in the deserts of the Middle East; however, some of them shed their leaves and their water output is thus reduced. The presence of plants with different life forms in desert plant communities results in a diversity of responses to drought as well as variable amounts of water output during the different seasons.

8.2.5.3
Growth Form Diversity

A distinction should be made between the term growth form and the term life form, as defined by Raunkiaer. According to Kassas and Imam (1959), growth form is distinct from life form, in that the latter refers to an inherent characteristic of a plant, while the former refers to a characteristic acquired by the plant under the prevailing environmental conditions. This distinction is made with respect to several species that grow as deciduous plants under unfavourable habitat conditions, e.g. during the summer drought, whereas under favourable conditions they are evergreens. Growth form presents a complete morphological and physiological picture of the plant as regards its response to climatic factors. The growth forms of desert plants include ephemerals, evergreen perennials, summer-deciduous perennials, winter-deciduous perennials and succulents. Plants in these categories vary in their response to the seasonality of the climate, especially temperature and rainfall. The desert plant community consists of different species with different growth forms. Each of these growth forms has a special response to the prevailing conditions, especially soil water. Such responses vary from one species to the other. The responses of a few species result in reduced water output when the water supply is low. There is always a balance between the amount of water available and water output. This balance keeps the community, with all its components, performing its functions satisfactorily.

Despite the very strict division between annuals and perennials, some authors have regarded plants recorded as "biennials" as either annuals or perennials. In a good rainy season, many perennials succeed in flowering and setting fruit in their first year. So even if they are killed by drought the following summer, they will have completed a full life cycle as potential annuals. This may be an important factor in their survival as a population. Haines (1951) described how the life cycles of different species are related to prevailing environmental conditions and showed how some perennials behave as potential annuals. From Haines' observations on the development of certain plants, it has been established that some perennial species, such as *Diplotaxis harra* and *Cleome africana* (reported as *C. arabica* in Haines' paper), flower regularly and profusely in their first season, but more often survive far longer. *Farsetia aegyptia* flowers early in the season, *Zilla spinosa* only flowers occasionally, and *Acacia tortilis* does not flower at all.

8.2.5.4
Physiological Diversity

Within many desert plant communities, there are numerous species with different photosynthetic pathways. The majority of desert plants communities include both C_3 and C_4 plants. CAM plants may also be present. The occurrence of plants with different photosynthetic pathways can be regarded as a means by which the community maintains positive water relationships. C_3 plants are usually active at times when the temperature is low (winter in the Middle East). They flourish during this period, and their foliage and vegetative growth are optimal (if there is any optimum in the desert!). They have a high rate of transpiration during this period. By means of different adaptations, the rate of transpiration decreases in the dry season. In contrast, C_4 plants, such as *Panicum turgidum* and *Pennisetum divisum*, flourish during the hot weather, which starts in April and persists until the end of summer. C_4 plants, such as *Haloxylon salicornicum* and other chenopods, prosper and flower at the end of the dry season. They are arido-active plants (Evenari et al. 1971), and withdrawal of water from the soil takes place during the hot season (in the Middle Eastern deserts this is the dry season). These C_4 plants possess high water efficiency and are more economical in their water use than are C_3 plants, so their water requirements are lower. Water loss is also limited when CAM plants are present and no extravagant withdrawal of water occurs. From the above discussion it can be seen that the diversity of photosynthetic pathways in the different components of many desert communities usually ensures a continuous supply of water throughout the year. Each component receives its share during a particular time of the year and does not interfere with the supply of others. This can be regarded as "ecological justice"!

It is not only the life history of a plant that may be related to its photosynthetic pathways, but also its season of activity. Table 4 shows the photosynthetic pathways of grasses in Egypt with their different growth forms and periods of activity.

Table 4. Photosynthetic pathways of Egyptian grass species with different growth forms and periods of activity. (After Batanouny et al. 1988)

Growth form and season of activity	C_3 species (*n*)	C_4 species (*n*)	Total (*n*)
Winter annuals	85	25	110
Perennials flourishing in winter	11	–	11
Perennials flourishing in spring or the whole year	9	38	47
Perennials flourishing in summer	–	35	35
Summer annuals	–	22	22
Total (*n*)	105	120	225

8.2.5.5
Intraseasonal Diversity

Although all the desert annuals appear after the rains, which fall from November to April, their germination and their vegetative and reproductive growth periods show temporal differences during the wet season. These are controlled mainly by day length and partly by temperature. In arid and semi-arid regions, waves of different species appear at different periods of the wet season. Those species appearing in the early winter (December–January) are C_3 annuals, which are adapted to the relatively low temperature prevailing in these months. Annuals appearing in the late winter (February) live till late spring (May to June); they are mainly C_4 plants (especially grasses and some succulent annuals). These waves of diversity are usually obvious in good rainy years. Thus the temporal distribution of annuals in the arid zone allows the various species to exploit the uppermost soil water (to a depth of ca 50 cm) at different times of the wet season. This is known as intraseasonal diversity. It assists in both the distribution of water and its withdrawal, as well as in the distribution of the accompanying fauna. Of these, the insects, which perform the function of pollination, are especially important to plants. Different insect species are also distributed temporally and thereby avoid competition.

8.2.5.6
Allelopathy in Desert Plants

Among desert plants, it is to be expected that there should be considerable competition for the limited water available. This results in the development of many competitive effects, including allelopathy, which are exemplified in those plants able to survive under harsh conditions (Harborne 1977). *Encelia farinosa* (Compositae), which grows in the Mojave Desert of central California, as described by Went (1942), provides an example of a shrub with bare patches of soil underneath its canopy and around it. In this climate zone, annuals do not flourish. By inhibiting the growth of annuals, the shrub thus ensures for itself the available moisture within a metre or so of its site of growth. Gray and Bonner (1948, cited in Harborne 1977) have shown that the fallen leaves contain a toxin that is not self-inhibitory but which engenders pronounced growth inhibition in many other plants.

Allelopathy among the plants of the Middle East deserts needs to be further investigated. Factors other than allelopathy help to control the development of new seedlings around some shrubs. One of these is the debris, trapped by the plants during the torrents that follow heavy rainfall. Some seeds must undergo scarification before germination can occur. For example, the seeds of certain leguminous plants, which have especially hard and impervious testa, require very thorough scarification. Thus, the seeds produced, for example by *Retama* or *Acacia* trees, do not germinate near the mother plant. If this were to happen, there would be an aggregation of individuals of the same species in a limited space. This, eventually, would create competition for water. These seeds do not, however, germinate in nature unless they have been dragged by torrents of water or blown by wind across rock and gravel surfaces, which causes the removal of part of the testa. After this treatment, the seeds are able to germinate, whenever external conditions are favourable. In other words, the opportunity for germination does not appear until the seeds have been separated from the mother plant. This is one of the many additional phenomena that enable desert plants to survive under harsh desert conditions.

8.2.6 Plasticity of Transpiration Rates in Different Microhabitats

Topographic differences have been shown in separate microclimatic and edaphic conditions in the various microhabitats of a desert wadi. Some plant species can grow in several different microhabitats. In order to see how far environmental conditions influence the transpiration rate of the same species in different habitats, Abdel Rahman and Batanouny (1965b) measured the transpiration rates of two species in different microhabitats. The diurnal fluctuations in the evaporation and transpiration rates of *Zilla spinosa* and *Zygophyllum coccineum* in three different microhabitats in the dry season are presented in Fig. 49. The data show how the daily rhythms of transpiration in both species vary from one habitat to the other. This evinces the plasticity of the transpiration response of desert plants depending on prevailing conditions. The early decline in the transpiration rate in the plateau and the first terrace represent an economy in the use of water during a time of high transpiration. In the shade, it may be a result of to a general decline in the evaporating factors.

Plants are naturally subjected to different environmental conditions in the desert, namely: (1) humid atmospheric and edaphic conditions that usually occur in the rainy season, (2) humid atmospheric conditions combined with soil drought, that prevails in years with late winter rains and a prolonged drought period extending through the early winter, and (3) combined soil and atmospheric drought prevailing in summer when the temperature is high, the air is dry, and no rain falls. Atmospheric drought and favourable soil moisture represent a fourth combination. This is not found in the Middle East, except when there is early rainfall, usually of the thunderstorm type, in October. The situation can be created artificially by irrigating plants in summer to increase the soil moisture to field capacity in the root zone. Under the above-mentioned conditions, Abdel Rahman and Batanouny (1965b) investigated the daily changes in the transpiration rates of *Pennisetum*

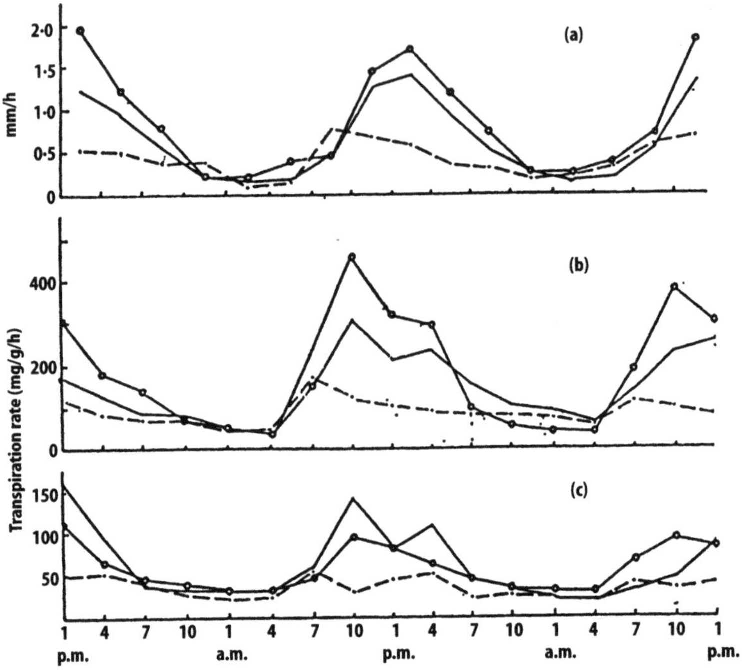

Fig. 49. Diurnal fluctuations in **a** the evaporation rate, and the transpiration rates of **b** *Zilla spinosa* and **c** *Zygophyllum coccineum* in the various microhabitats in wadi Hoff, the Egypto-Arabian Desert (July 1960). *Open circles* Plateau; *smooth line* first terrace; *dashed line* shade. (After Batanouny 1963)

divisum (P. dichotomum) and *Zilla spinsa* (Fig. 43). They showed that desert plants, when supplied with water during the summer months, have very high transpiration rates. The attainment of high transpiration rates under these conditions is caused by the high evaporation rate. This, in turn, is due to favourable conditions for water absorption, namely abundant supply and high soil temperature. Comparison of the maximum and mean transpiration and evaporation rates for the two species under different conditions (Table 5) shows that, when atmospheric drought was associated with soil drought, i.e. in the July experiment in which irrigation was not applied, maximum transpiration rates in the two species were much lower than in the irrigated plants, being 455 mg g^{-1} h^{-1} in *Pennisetum*, and 307 mg g^{-1} h^{-1} in *Zilla*. The means transpiration rates under natural conditions were 210 mg g^{-1} h^{-1} in *Pennisetum* and 140 mg g^{-1} h^{-1} in *Zilla*. In the watered plants, the mean rates were 275 mg g^{-1} h^{-1} in *Pennisetum* and 242 mg g^{-1} h^{-1} in *Zilla* This indicates that soil drought in the rainless season restricted the transpiration rate to a low level. In other words, well-watered desert plants transpire more

Table 5. Comparison of the maximum and mean transpiration rates of *Pennisetum divisum* and *Zilla spinosa* and the evaporation rates under different environmental conditions. (After Abdel Rahman and Batanouny 1965b)

	Transpiration rate (mg/g/h)				Evaporation rate (mm/h)	
	Pennisetum		*Zilla*			
	Maximum	Mean	Maximum	Mean	Maximum	Mean
July, watered	711	275	617	242	1.40	0.65
July, dry	435	210	307	140	1.40	0.65
March	295	149	273	151	0.66	0.24
December	196	104	270	137	0.73	0.32

freely than non-watered ones. One can see that desert plants are not characterised by low transpiration rates, but rather by their ability to lower the rate of transpiration under soil moisture deficits.

CHAPTER 9

Adaptational Traits That Resist and/or Tolerate Drought

9.1 Introduction

The terms "tolerance" and "resistance" are usually employed interchangeably and loosely. Before we discuss the traits that help desert plants to tolerate and/or resist drought, it is necessary to evaluate these two terms within the present context. Drought tolerance means the degree to which a plant can successfully withstand the effect of drought. This occurs as a result of the innate characteristics of the plant. It may mean that a plant does not exhibit modifications or changes in its structure or function to do so. This can be considered to be a passive response. Resistance has another meaning – it is an active response to the factor affecting the plant. Drought resistance can be understood to mean that the plant remains unharmed by the drought through various adaptations and traits. These traits can be modified and/or maximised according to the degree of severity, and the duration and the length of the drought. In order to live under desert conditions, plants in general possess the characteristics of both tolerance and resistance to drought. It seems probable that resistance traits come into action when tolerance traits have reached full capacity. In other words, when a plant is subjected to stress, its tolerance characteristics oppose this stress. As the stress increases, the activity of tolerance traits continues to rise in opposition. At a certain threshold, the tolerance capacity becomes insufficient to withstand the stress. At this stage, the resistance traits are initiated. The response of these traits can be short-term or long-term, according to the severity and duration of the stress. Every species has its own potential for tolerating and resisting drought. When grown under different environmental conditions, different individuals of the same species exhibit different degrees of resistance. The drier the conditions, the greater the degree of resistance manifested. This is due to the fact that desert plants are flexible as regards their response to drought. They are endowed with many adaptational traits and characteristics that help them to tolerate and resist drought.

Under desert conditions, plants are usually subjected to water stress. The term "stress" is in common use in the plant physiology literature but its meaning has at times been controversial (Kramer 1980; Levitt 1980). The numerous physiological responses of plants to water deficit usually vary with the severity as well as the duration of the stress. Only the most sensitive processes are altered by mild stress. As the stress increases, however, the changes in the plant intensify (this is resistance) and additional processes become affected in accordance with their relative sensitivities (this is tolerance).

Among the terms used to describe water conditions in the plant body is "water status". It is a vague but nevertheless useful term (Passioura 1982). Water status has no units that refer to any measure of the general state of a plant in relation to water; rather, the term describes a relative state. It may refer to a plant's water potential, its relative water content, or its turgor (Passioura 1982).

9.2 Physiological Adaptations

It has been claimed that water deficiency influences the behaviour of a plant at all levels of organisation: metabolism, physiology and gross morphology are all affected. The role of growth morphology as an adaptive means of controlling water loss was discussed in a previous chapter. It is clear that such means are a powerful method of influencing the water economy of a plant. In the following pages, we shall discuss the different physiological traits of desert plants that help them to tolerate and/or resist the conditions of drought under which they live.

9.2.1 Stomatal Control

Levitt (1976) stated that stomata fulfill three major functions in the physiology of plants: (1) They restrict the entrance of CO_2 into the green leaf to a rate sufficient to support adequate photosynthesis for normal growth and development. They have become adapted to this function by opening in response to low concentrations of CO_2 inside the leaf. (2) They permit the entrance of O_2 into the leaf at a rate rapid enough to support aerobic respiration sufficient to provide for metabolic needs. They have become adapted to this function by opening in response to low O_2 concentrations inside the leaf. (3) If these were the only functions of the stomata they would fulfill them by remaining constantly open. Accompanying the uptake of CO_2

and O_2, however, there must also be an increased loss of water vapour, as long as the water potential of the external atmosphere is below that of the leaf. Levitt (1976) added that, when loss of water approaches the point of stress injury, the third function of the stomata is to close, reducing the total rate of water loss to that caused by cuticular transpiration – a small fraction of stomatal transpiration. Stomata have become adapted to this function by means of a guard-cell structure, which leads to their closure when loss of water from the cells lowers the turgor pressure. Recent evidence has revealed another adaptation that permits stomata to close even before evaporative loss of water is sufficient to lower the guard cell turgor (Lange et al. 1971). In this case, closure must be due to osmotic loss of water, again followed by loss of guard-cell turgor.

The physiology of desiccated plants has received considerable attention, especially when with regard to the behaviour of the stomata. Stomatal conductance depends on the water status of a plant. This in turn depends on the water status of the soil, on the hydraulic resistance experienced by the transpiration stream, on the transpiration rate, and on the osmotic pressure of the symplast.

Stomatal responses are the principal mechanism for short-term carbon gain and water loss in plants. This regulation operates within specific ranges and depends on the naturally variable conditions of the habitat. For arid conditions, Stocker (1956) distinguished between daily patterns of gas exchange, namely one-peaked and two-peaked, and he associated these and various intermediate forms with the morphological adaptations of different species as well as with environmental conditions such as light, humidity and water stress.

Plants sometimes exhibit one-peaked patterns of gas exchange at very low rates of CO_2 uptake and water vapour loss resulting from long-term drought (Stocker 1956). With more extreme conditions of soil drought a short morning peak of CO_2 uptake was observed, followed by CO_2 losses in the light (Schulze et al. 1980). In this case, the transpiration rate was low and mainly due to cuticular water loss. Under natural environmental conditions one-peaked patterns at low rates of CO_2 uptake may be due to high daytime temperatures, and leaf conductance may be low due to the effects of low air humidity in addition to severe soil water stress. The types of transpiration curves, as affected by environmental conditions, have been investigated by Batanouny (1974). They were discussed in Chapter 8 (this Vol.). The variation in the transpiration curves reflects the activity of the stomata and their response to the water status of the plant body.

9.2.2 Bound Water

It has been observed that the drying of cut plant shoots in the air gradually slows down, and water loss decreases until it reaches zero. In dry air, the shoots of cut plants still retain some water in their tissues. This water can only be eliminated by oven drying. The percentage of water retained in the air-dried shoots of desert plants is higher than in mesophytes. Migahid (1938) stated that plants of drier habitats possess an enhanced and elaborate mechanism of defense against dehydration. This is manifested by both increased hydrophilicity and increased stability of the plant tissues. He claimed that increased hydrophilicity is demonstrated by the greater resistance to transpiration and a higher proportion of "bound" (retained) to "free" (lost by transpiration) water in xerophytes than in mesophytes. Increased stability is shown by the greater readiness with which the plants of drier habitats absorb water from the saturated atmosphere.

The length of time during which transpiration decreases to zero during air-drying differs in different species. Migahid (1938) found that this time extends to 22 days in *Vinca rosea*, 16 in *Zygophyllum*, 11 in *Peganum harmala*, 14 in *Withania somnifera* and 6 in both *Pulicaria crispa* and *Stachys aegyptiaca*.

When equilibrium is established between the vapour pressure outside and inside the plant, some water is retained in the air-dry shoots. This amount has been found by Migahid (1938) to be equivalent to 2.2 % of the total water content in *Vinca*, 2.9 % in *Withania*, 4.4 % in *Zygophyllum*, 7 % in *Peganum*, 8.9 % in *Stachys* and 11 % in *Pulicaria*. It is thus apparent that xerophytes are more capable than mesophytes of binding water.

Carr and Gaff (1961) and Gaff and Carr (1961) were the first in recent times to suggest that the water in cell walls – apoplasmic water – could provide a reservoir, protecting cells from water loss by acting as a buffer for symplasmic water. Cutler et al. (1977) suggested that high apoplasmic water content is a common feature of drought-adapted plants. However, it is difficult to draw any correlations between the amount of apoplasmic water and drought adaptation (Tyree and Jarvis 1982). Removal of the apoplasmic water from the cell walls requires high tension. This implies that the apoplasmic water is retained in the walls and will be lost only under extreme drought conditions.

9.2.3 Allocation of Resources

Desert plants have been found to allocate relatively fewer resources to their vegetative organs than to the reproductive ones (Boaz et al. 1994). This is consistent with the hypothesis of Noy-Meir (1986), that the first shift to the reproductive stage in desert conditions is triggered by water stress, and with Kozlowski and Wiegert (1986), who proposed a similar response for plants growing under conditions of short seasons and high mortality rates. The pattern is also compatible with previously proposed models (Cohen 1971; Wilson 1983) of maximisation of reproduction by an early shift from the vegetative to the reproductive stage.

As expected, drought accelerates the transition from the vegetative to the reproductive phase, and a large proportion of the biomass is then allocated to reproduction. Nonetheless, vegetative growth may continue during the reproductive phase even when a large proportion of biomass is allocated to reproduction. Cohen (1976) has argued that, if vegetative growth is an added advantage to survival or reproductive yield, a prolonged phase of both vegetative and reproductive growth can be expected to take place between the solely vegetative and solely reproductive phases.

When various individuals of the same annual species grow in runnels crossing the slopes on the desert hills, vegetative growth is different among different individuals. This variation is controlled mainly by the width of the runnel which, in turn, affects soil depth and water supply to the site. In other words, the water supply through runoff and its preservation by deep or shallow soil vary in different sites along the runnel. In the narrow parts of the runnel, where the soil is shallow, plants have small vegetative bodies and produce flowers earlier than they do in sites with deeper soil and a better water supply. In the latter, vegetative growth is more vigorous and flowering is late. In dry sites, some annual species are able to complete their life cycles within 2 weeks. This phenomenon can be observed along small runnels or the edges of depressions in the desert. The wider parts of the runnel and the centers of the depressions usually support plants with better vegetative growth and later flowering. The allocation of resources to the reproductive organs under dry conditions is of considerable significance to the life of plants in desert environments. If the plant does not shift its resources toward reproduction at the early stages of drought, it may die without producing seeds. In that case a potential seed population will be lost forever. But the allocation of resources to reproductive organs allows annuals to produce more seeds, thereby compensating for those lost from the seed bank. Hence, this phenomenon plays a role in survival - otherwise, the germination of seeds in subsequent years, during which the life cycle is not completed, would lead to extinction.

9.2.4
The Accumulation of Proline and Other Amino Acids

Water stress induces certain biochemical processes that may be of ecological importance under conditions of water deficit. Among the characteristic responses to water stress is accumulation of the amino acid proline. This appears to be an important stress metabolite and it accumulates during relatively mild water stress (water potentials of −1.0 MPa; Crawford 1989). The phenomenon appears to be a general one, as greater amounts of proline have been recorded in wilted than in turgid leaves of Bermuda grass (Barnett and Naylor 1966), *Citrus* (Chen et al. 1964), wheat (Prusakova 1960), ladino clover (Routley 1966) and other plants (Stewart et al. 1966). Batanouny and Ebeid (1981) found that different desert perennial species (*Zygophyllum qatarense* and *Pulicaria crispa* (syn. *Francoeria crispa*) accumulate different amounts of proline under the same environmental conditions. They also showed differences in the diurnal changes in proline content. Batanouny and Ebeid (1981) concluded that there is causal relationship between proline increase and water content decrease (Figs. 50, 51). The curve showing diurnal changes in the proline content of watered and naturally growing *Zygo-*

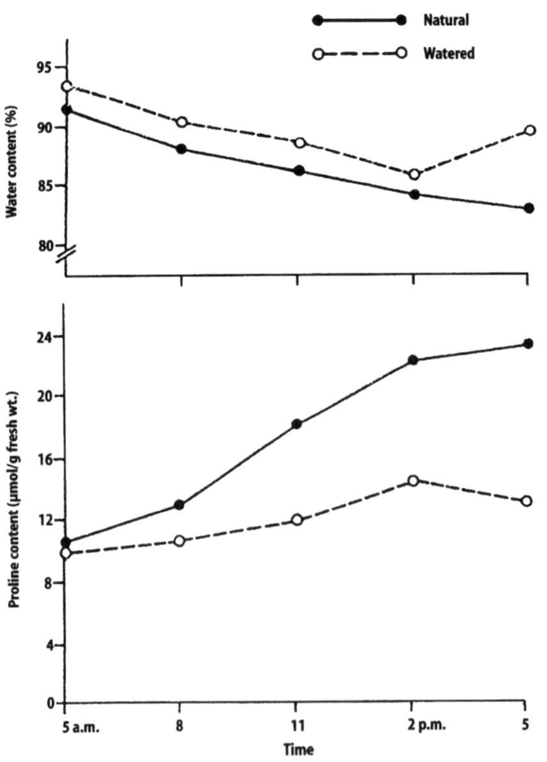

Fig. 50. Diurnal changes of proline and water contents of natural and watered *Zygophyllum quatarense* plants under desert conditions. Qatar, Arabian Gulf. (After Batanouny and Ebeid 1981)

Fig. 51. Diurnal changes of proline and water contents of natural and watered *Pulicari crispa* (syn. *Francoeuria crispa*) plants under desert conditions. Qatar, Arabian Gulf. (After Batanouny and Ebeid 1981)

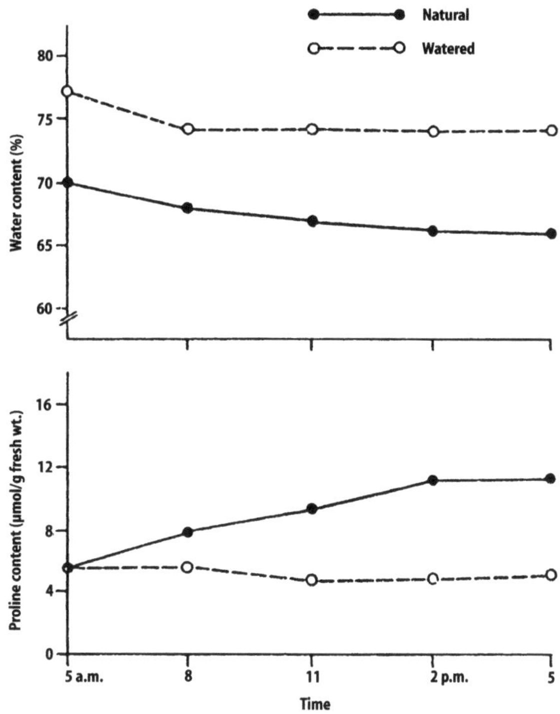

phyllum qatarense is a mirror image of the curve showing changes in water content (Fig. 50).

In a study on proline accumulation in *Retama raetam* (Batanouny et al. 1985), low proline content was observed before sunrise and then increased considerably as the day progressed, decreasing to low levels after sunset. The same authors showed that the curves plotting the diurnal changes in proline content were parallel to those of water saturation deficit. They found that cut *Retama* branches starved of water and exposed to natural desert conditions accumulated proline in their tissues. Some 2 min after cutting, the proline content was 63.41 µmol/g and increased to 88.42 µmol/g after 5 h. This rise can be explained by the increased water stress resulting from the decrease in water content from 46.9 to 31.7 %.

Many interpretations of the importance of proline accumulation have been presented. It has been claimed that proline accumulation might be involved in osmoregulation. Osmoregulation refers generally to the accumulation of solutes over and above that caused by fluctuating water levels per se (Crawford 1989). One might therefore decide that an increase in the proline level cannot account for any substantial increase in the osmotic potential of the cells of higher plants (see Kluge 1976; Schobert 1979, 1980).

Singh et al. (1973a,b) argue that the major importance of proline accumulation is a hardening effect rather than osmoregulation. There are also suggestions that proline may provide a convenient source of energy and nitrogen in immediate post-stress metabolism (Stewart 1973). Schobert (1980) suggested that proline accumulation regulates cell water. She postulated that the accumulated proline associates with hydrophobic areas of biopolymers, thus converting hydrophobic groups into hydrophilic ones. By this mechanism, the solubility of biopolymers, which is impaired as a result of hypersomatic stress, is increased and the problem of hydrophobic hydration is minimised.

Though proline accumulates in the tissues of different plants at various ecological amplitudes in response to water stress, the amount of proline cannot be taken as a diagnostic feature of plants of different ecological groups. Moreover, the proline-accumulating potential should not be taken to be a positive index of drought resistance. This is true even in cultivars of the same species. Hanson et al. (1977) obtained data on two contrasting barley cultivars that suggest that reports of a simple positive correlation between proline-accumulating potential and drought resistance in barley may be in error.

9.2.5 Photosynthetic Pathways

Different photosynthetic pathways have been recognised in the various components of a desert plant community. Under desert conditions, one can expect that these difference are important with respect to plant-water relations. In previous chapters, this has been shown to be the case.

9.2.5.1 *Traits of C_4 Plants*

The C_4 pathway of photosynthesis is always associated with Kranz anatomy in the leaves, in one or more of its complex forms. On the basis of anatomical observations, Volkens (1887), Haberlandt (1884) and Rhoades and Corvalho (1944) speculated on the possibility of a "division of labour" during photosynthesis in plants that displayed the Kranz arrangement of photosynthetic tissues. The vascular bundles of C_4 grasses are surrounded by special photosynthetic tissue (Fig. 52; see also the anatomy of the grasses given by Volkens, Fig. 54).

Plants with a C_4 pathway discriminate less against the heavy C isotope during carboxylation than do C_3 plants (Bender 1968). The carbon isotope discrimination properties of plants are readily evaluated by measurement of the ratio of ^{13}C to ^{12}C in CO_2 produced by the combustion of dried plant

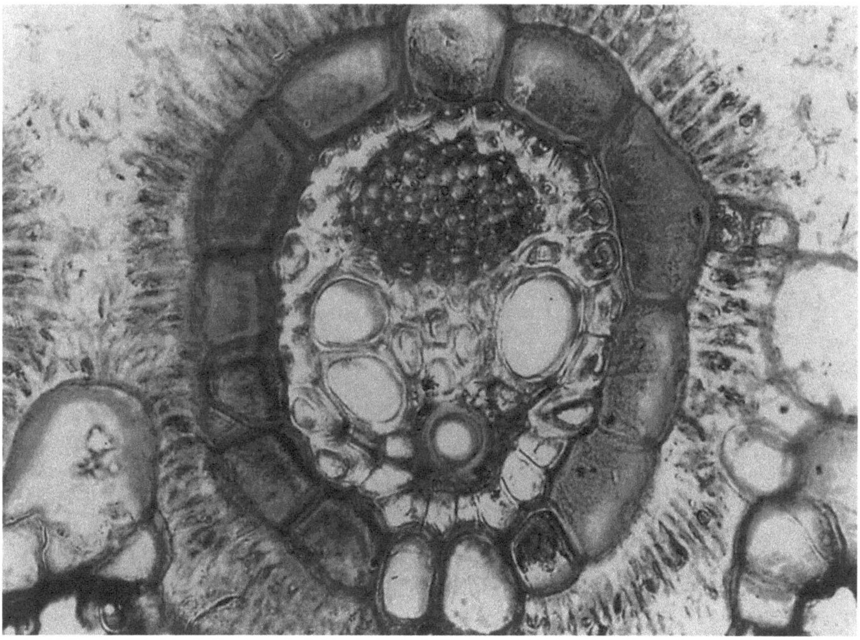

Fig. 52. Transverse section in *Sporopolus arabicus* leaf showing the special arrangement of the photosynthetic tissues around the vascular bundle

material. The ratio is expressed as a $\delta^{13}C$ value, in parts per thousand, after reference to a limestone standard:

$$\delta^{13}C(‰) = \left[\frac{^{13}C/^{12}C \text{ sample}}{^{13}C/^{12}C \text{ standard}} - 1\right] \times 10^3$$

With access to an appropriate mass spectrophotometer, the method can rapidly and reliably distinguish between terrestrial C_3 and C_4 plants (Osmond et al. 1982). The reported $\delta^{13}C$values of C_4 plants range between -6 and -17 %.

Under identical conditions of leaf temperature and leaf-air vapour pressure differences, the lower stomatal conductance of C_4 plants is responsible for substantially lower transpiration from the leaves than measured in C_3 plants (Osmond et al. 1982). The transpiration/assimilation ratio is greater in C_3 plants than in C_4 plants: C_4 plants have a higher water use efficiency (Osmond et al. 1982). However, this cannot be taken to imply that the lower transpiration ratio of C_4 photosynthesis necessarily results in more water-efficient carbon acquisition under natural conditions. The water economy of crassulacean acid metabolism (CAM) plants is superior to that of

Fig. 53. *Euphorbia cactus* Ehrenberg growing in Yemen, South Arabia

both C_3 and C_4 plants, principally because CAM plants assimilate the bulk of their carbon at night, when leaf temperature and water vapour pressure difference are low. Some succulent *Euphorbia* species that grow in the deserts of the Middle East are CAM plants. They are found in deserts where high temperature is a common feature of the climate (Batanouny et al. 1991). *Euphorbia cactus*, a CAM plant, is among the common plants growing in southern Arabia (Fig. 53).

One should take into consideration the fact that there is no a priori reason to assume that the less expensive water economy of C_4 plants makes them any more tolerant of low tissue potentials.

The distribution of C_4 plants seems to be affected more by temperature than by water (Batanouny et al. 1988, 1991).

9.2.6 Osmotic Potential

The term "osmoregulation" characterises the "effort" of a plant to minimise differences in water potential between the plant body and its environment by creating a suitable internal osmotic potential. For example, if the water potential of the environment decreases, i.e. becomes more negative (for example due to a high water-vapour saturation deficit of the atmosphere, decreasing water content of the soil or an increasing solute concentration in the aquatic environment), plants exhibiting osmoregulation can lower (make more negative) their own internal water potential by the synthesis and accumulation of osmotically active substances, and vice versa. By this mechanism, plants are able not only to prevent loss of water to an environment having large negative water potential, but also to take up water from such an environment, a fact that enables them to balance the water deficits of their cells.

The osmotic pressure of cell sap is nearly equal to the suction force or diffusion pressure deficit (DPD), since the cell rarely if ever reaches a state of turgidity under arid conditions. For this reason, the osmotic pressure of desert plants is a good criterion of the ability of these plants to withdraw water.

Abdel Rahman and Batanouny (1964b) found that there is a wide range of osmotic pressure values in the sap of desert plants. In their study of plants growing in different habitats in wadi Hoff, in the Egyptian Arabian Desert, osmotic pressures varied from 10.4 to 35.6 atm. The authors found that species such as *Zygophyllum coccineum* (28.1 atm), *Farsetia aegyptia* (20.7 atm), *Achillea fragrantissima* (20.7 atm) and *Lycium shawii* (syn. *L. arabicum*) (20 atm) have relatively higher osmotic pressure values than other perennials. It is interesting to note that the osmotic pressure values of annuals are low, e.g. *Trigonella stellata* (13.2 atm), *Diplotaxis acris* (13.5 atm). However, some perennials, such as *Pennisetum divisum* (syn. *P. dichotomum*) (12.9 atm), *Zilla spinosa* (14.2 atm) and *Lavandula coronopifolia* (11.2 atm), also have low values. A few annuals have high osmotic pressures, e.g. the succulent *Zygophyllum simplex* has a value of 28.9 atm.

Study of the daily variation in osmotic pressure values revealed that maximum pressure is usually attained in the late afternoon or early evening, while the minimum is attained after midnight (Abdel Rahman and Batanouny 1964b).

Seasonal variations in the osmotic pressure of the different species were also noted. The widest range occurred in shallow-rooted plants and in succulent plants. The maximum osmotic pressure for many of the desert perennials studied by Abdel Rahman and Batanouny (1964b) was attained at the end of the dry period as a result of the depletion of water from the soil and the low water content of the plant tissues.

It is interesting that the same species growing in different microhabitats exhibit different osmotic pressure values. In the microhabitats of wadi Hoff, the highest osmotic pressure for *Zygophyllum coccineum* was recorded on the plateau, with the evaporative power of the atmosphere reaching a mean annual value as high as 31.8 atm, while it was 28.1 atm in the wadi bed and only 25.4 atm in shaded areas (Abdel Rahman and Batanouny (1964b). In earlier observations, Korstian (1924) noted that sun plants had higher cell sap densities than shade plants. Marsh (1940) emphasised the influence of direct sunlight on the development of relatively high osmotic pressure. This agrees with the results of Oppenheimer (1932) and Migahid (1962), that the same species has a higher osmotic pressure when growing in a dry, exposed locality than when growing in a moist, shaded one.

Irrigation of desert perennial plants, such as *Zilla spinosa* and *Pennisetum divisum* (syn. *P. dichotomum*), in summer resulted in a considerable reduction of the osmotic pressure values of their cell sap. The irrigated specimens of *Zilla spinosa* had a mean osmotic pressure of 13.9 atm, while those growing under natural conditions, spaced a few metres away, had a mean osmotic pressure of 15.6 atm (Abdel Rahman and Batanouny 1964b).

Abdel Rahman and Batanouny(1964b) also found that there is no definite correlation between the osmotic pressure and the transpiration rate. In the case of *Achillea fragrantissima*, the osmotic pressure is 20.7 atm and the transpiration rate is as high as in some annuals. By contrast, the transpiration rates of *Zilla spinosa* and *Pennisetum divisum* are low and their osmotic pressures are also low compared with those of other perennials.

9.2.7 Protoplasmic Characteristics

Over at least a 4-h time period under summer desert conditions, the temperature in the sun may be above 50 °C; solar radiation is very intensive, and the water contents of plants are to a great extent reduced. The plant is anchored to the ground and exposed to drastic conditions. The normal survival mechanisms discussed in the present book are not sufficient to secure plant life under such conditions. One assumes that the protoplasmic components of the cells of perennial desert plants active in summer have specialised features that distinguish the respective plants from those of mesic regions. However, one cannot say which component of the protoplasm is most effective in this respect, and which is destroyed by the extreme conditions prevailing in the desert.

Bewley and Krochko (1982) stated that "Water deficits can affect cellular processes in one or more of the following ways; (1) through a decrease in turgor pressure; (2) by a reduction of the chemical potential or activity of water in cells; (3) by changes in spatial relations in membranes and organelles resulting from decrease in volume of the protoplasm; (4) by effects of increase in solute concentrations caused by losses of water; (5) by alteration of structure of macromolecules resulting from dehydration (Hsiao 1973)."

According to Stocker (1960; reviewed by Levitt 1972; Mahmoud 1965; Parker 1968), the tensions occasioned by desiccation act not so much on the surface layer of protoplasm but rather within the internal protoplasm. He believed that the protoplasm consisted of a latticework of threadlike proteins interspersed with water, salts, sugars, and various proteins. The initial cellular responses to water stress, which he termed "the reaction phase", result in an unequal hydration capacity of different parts of this protein framework. This leads to disorganisation and loosening of the protein latticework as water is withdrawn, resulting in a reduction in the viscosity of the protoplasm and an increase in its permeability. If drying is gradual enough, the cell subsequently enters a phase of restitution or restoration during which protoplasmic bonds are strengthened, a reorganisation of the protein latticework occurs, and viscosity increases. These changes result in overcompensation of the drought reaction and the plant enters a hardened condition. During rapid drying, the plant is prevented from completing the restitution phase, consequently there is an increase in drought-associated injury. In plants that are able to recover from air-drying, Stocker postulated a reactivation phase following remoistening that repairs the damage caused by desiccation.

Stocker's view that the secret of drought-tolerance would be found in the molecular organisation of the protoplasm was a significant departure from the theory proposed by Iljin, in which cell morphology was considered to be of primary importance. However, Iljin's hypothesis (1935, 1953) regarding the destruction of protoplasm by dehydration and rehydration, which results in the death of the tissues under water stress, seems at least partially confirmed by Viera da Silva's (1976) study.

9.2.8
Response to Heat Stress

Over-heating is one of the main problems faced by desert plants. Unable to cool their leaves by increased evapotranspiration, they undergo a range of morphological changes that reduce heat loads as well as of physiological adaptations that permit tolerance of higher temperatures. Overheating is avoided by: (1) increasing the surface area for heat dissipation; (2) protecting the plant surface from direct sunlight; and (3) increasing the reflectance of the surface.

All these effects can be accomplished by covering the surface of the plant with a dense felt of hairs. These hairs also produce a boundary layer of still air next to the stomata that traps moisture and reduces evapotranspiration. Hairs may also have a role in defense against herbivorous insects (Cloudsley-Thompson 1996). Not all xerophytic plants are hairy, however, and some desert species have completely hairless, smooth surfaces. In these cases, water loss is restricted by a thick covering of wax, or by modifications of the walls of the epidermal cells (Levitt 1972).

Desert trees have small, finely divided leaves, or sometimes none at all. Leaves heat up in the desert sun. Structures lose heat by convection most rapidly at their edges; the more edges there are, the cooler will be the leaf and the lower the water loss. Small size means a considerable portion of each leaf is occupied by its edge.

The leaf temperature may be lower (under-temperature types) or higher (over-temperature types) than the air temperature (Lange 1965a). The temperature of a leaf depends on many factors that, in turn, are affected by the external morphology and anatomy of the leaf as well as by external conditions existing at the time of the measurement (Lange 1965a). *Citrullus colocynthis* is a typical example of an under-temperature plant (Lange (1965). The tissue temperature of the leaf remained below the air temperature with differences of 10–12 °C; this was explained by the strong cooling engendered by transpiration. In a cut leaf with a decreasing transpiration rate, the leaf temperature increased rapidly. It reached the air temperature level and even passed it. Lange (1965a)also showed that a species of *Zygophyllum* had a leaf temperature 10 °C above the surrounding ambient air temperature (30 °C, and discussed the factors considered by Huber (1956) and Raschke (1956) to be most important in determining leaf temperature. These are radiation balance (i.e. the difference between adsorbed radiation and long-wavelength radiation), transpiration rate, the latent heat of vaporisation at a particular leaf temperature, convection heat exchange with the environment, chemical energy exchange (which was relatively insignificant), and heat exchange due to water transport. These are basically the same factors as those presented by Gates et al. (1968), although they also emphasised leaf width and orientation.

The thermal environment in deserts is determined by season, by abiotic factors such as temperature, radiation, wind, and moisture, and by biotically controlled factors such as shrub canopy growth. Plant microenvironments may have temperatures exceeding 60 °C with high solar radiation input, or temperatures not much above 20 °C with low solar radiation. Each site demands a different physiological response on the part of the plants that occupy it. Plant temperatures, which do not always follow air temperatures, are dependent on the plant's morphology, and its exposure, orientation, and ability both to prevent heat absorption by reflection and to dissipate heat through transpiration and convection.

9.2.9
Dew Absorption: Foliar Uptake of Water

Much of the literature on the absorption of water through the leaves and stems of vascular plants has focused on the possible significance of this pathway in arid environments. The suggestion that foliar water uptake might be important in desert plants was first made by Volkens (1887) in his studies of the physiological anatomy of the flora in the desert of Egypt. Until the present time, Egyptian scientists have referred to the absorbing hairs of *Diplotaxis harra*, as cited by Volkens. However, there is a great need to validate this assumption using new methods. This subject was also investigated at the Desert Botanical Laboratory at Tucson, Arizona, in the early part of the last century (Lloyed 1905; Cannon 1906; Spaulding 1905a, b). These studies, utilising a variety of simple experimental techniques to determine whether water up-take occurred in moistened leaves or stems, concluded that species with mesophytic leaves readily absorbed moisture, but the results with xerophytes and succulents were variable. In summarising the most extensive experimental work, Spaulding (1905b) concluded that foliar absorption was so slow that it was insignificant in comparison with root uptake.

Akhtar and Shaukat (1976, 1979) suggested that the utilisation of dew took place in three species of plants in Pakistan because they survived in soil moisture regimes below the permanent wilting point (−1.5 MPa). No critical demonstration of foliar uptake was included, however, nor was any suggestion made regarding the physiological tolerance of non-exposed plants to water stress.

Waisel (1958) carried out a series of experiments on the absorption of dew by desert and Mediterranean plants in Israel. He found that dew absorption was variable in different species, but in no case were there significant amounts of uptake and redistribution. He also noted that the rate of dew absorption increased as soil humidity decreased. Other studies carried out in Israel with *Tamarix* have suggested that the presence of dew and water vapour around the foliage is very important for the success of plantations of these trees (Gindel 1964). Further east, in the Asiatic desert belt, Sveshnikova (1972) described water vapor absorption by leaves and stems of a variety of plants from the Kara-Kum desert region of Turkestan. Again, however, unambiguous foliar uptake has not been demonstrated. Only two recent studies (Stark and Love 1969; Babu and Went 1978) have reported ecologically significant levels of water uptake through vascular plant leaves. The studies were done in North American deserts, and neither is supported by strong data.

Perhaps the most important benefit of dew under arid conditions is that it can provide a reasonable supply of water through the superficial roots (Chap. 8). The effect of dew in decreasing transpiration is especially important during the early morning.

Water uptake from the atmosphere is generally restricted to poikilohydric plants, since the cuticle of homeohydric plants, which is effective in restricting water loss, is also an effective barrier again water uptake. The hydration of vascular plants by dew may occur to a small extent; the topic is still controversial (e.g. Slatyer 1956; Kerr and Beardsell 1975). Despite the large existing body of literature suggesting foliar uptake of water by vascular plants, many more studies are needed to assess the physiological significance and ecological importance of direct foliar uptake of dew, precipitation and/or water vapor. In virtually every case, reports of the direct foliar uptake of water by vascular plants lack sufficient data to meet the necessary criteria for critical and unambiguous acceptance.

9.3 Anatomical Adaptations

9.3.1 Anatomical Features

A turning point in the botanical studies of the deserts of Egypt and those of the Middle East was the valuable study of Volkens (1887), already mentioned above. In 1884 the young German botanist Georg Ludwig August Volkens (born 13 July 1855 in Berlin) travelled to Egypt, where he stayed for 10 months. His studies in the Egyptian-Arabian desert investigated the relationship between the anatomical structure and the physiological function of plants. These, he believed, would be more obvious in the desert than elsewhere because of the overriding importance of water in the desert. He was the first botanist to study the relationships between the prevailing environmental conditions in the Egyptian desert and the structure of the indigenous plants. In 1887 he published his observations under the title: "*Die Flora der Aegyptisch-Arabischen Wüste, auf Grundlage Anatomisch-Physiologischer Forschungen*". He described certain anatomical structures in desert plants. In his opinion, these had evolved as functional adaptations to desert conditions. It seems proper to present here some of Volkens' beautiful drawings of the anatomy of some plants in the desert of Egypt, which were published in his book in 1887. Figure 54 shows plate no. XVIII in the treatise of Volkens. The drawings highlight the anatomical features of many desert grasses in Egypt. These species are widespread in the deserts of the Middle East and demonstrate the richness of diversity in the shapes, and consequently in their associated functions. It is to be noted that Volkens called attention to the role of hairs in absorbing the atmospheric humidity; he referred to the absorbing hairs of *Diplotaxis harra*. Since that time, discussion of the functional value of anatomical structure to desert plants has continued unabated. Evenari et al. (1971) raised many questions in this respect: Are there any anatomical structures specific to desert plants? What is their ecological value, if any? How are they related to the desert environment? Even today, we have no clear-cut answers to these questions. Our present understanding of the anatomical features of aboveground parts,

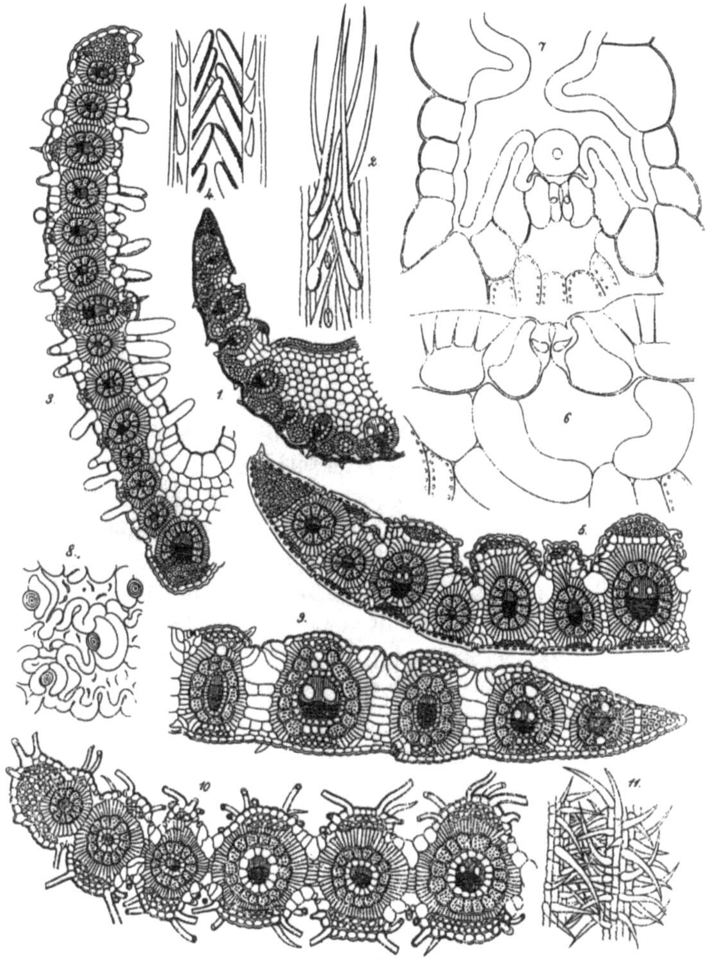

Fig. 54. The anatomical features of desert grasses in a plate drawn by Volkens. (After Volkens 1887)

especially structures that photosynthesise, is far better than that of the belowground parts.

Plants growing under desert conditions usually develop structural characteristics that are adaptations to arid habitats. It has been shown that the surface area of the leaves is reduced in desert plants, a feature correlated with the reduction of water loss. One of the most obvious features of the leaves of desert plants is the small surface-to-volume ratio. Fahn (1989) stated that reduction of the surface is accompanied by certain changes in the internal structures of leaves, e.g. reduction in cell size, increase in the thickness of the cell walls, greater density both of the vascular system and of the stomata and increased development of palisade tissue at the expense of spongy tissue.

Leaves of desert plants are often covered with trichomes. One cannot, however, claim that these play a role in reducing water loss. The trichomes themselves lose water through transpiration. Perhaps the dead trichomes serve as a protective layer that helps in reducing water loss and that protects the photosynthesising plant organs from insolation. At least dead trichomes reduce the temperature of exposed organs. Hairs or scales, as well as epicuticular wax and resin, may play a role in reducing insolation and in reflecting light (Johnson 1975; Ehleringer et al. 1976; Hartmann 1979; Heide-Jorgensen 1980; Fahn 1989).

Involution of the leaves, which is especially typical of grasses, is a characteristic of xerophytes. This feature is brought about by the action of bulbiform cells and/or other epidermal and mesophyllic elements that may be parenchymatous or sclerenchymatous (Shields 1950).

The well-known adaptations of roots to desert conditions include the succulence of the root system (e.g. in *Citrullus colocynthis* and *Hyoscyamus muticus*), the development of thick bark, sclerification of the cortical cells, and isolation of the vascular cylinder by formation of periderm or by necrosis of the cortical parenchyma (see Fahn 1989). Fahn (1989) discussed the role of well-developed xylem tissues, which help rapid conduction when water is available. The case *of Retama reatam* is given as an example of the presence of vertical roots, diageotropic roots, and horizontal roots. The xylem vessels of vertical roots are narrower and shorter than those of diageotropic roots and even more so than those of horizontal roots. Moreover, in the latter type of root a gradient in the length and width of the vessels exists from the distal ends of the roots to the part where they are attached to the vertical roots. The advantage of this feature may be its compensating effect on what would otherwise be a much more pronounced gradient in suction pressure from the stem toward the distal parts of the roots. Both the greater width and length of the vessels in the distal parts of horizontal roots presumably ensure a more efficient flow of water in very long, horizontal roots (which may reach 10 m). This feature is doubtless of great importance as the horizontal roots take up water from the upper layers, which in sandy soils drain rapidly. The vertical roots have xylem vessels with a narrow diameter whereas the vessels of the horizontal roots are wider (Fahn and Cutler 1992; Danin 1996). This appears to be an adaptation of the root system for the efficient conduction of water from the surface layers of the soil, which are often wetted only to a limited depth.

Two interesting features have been observed in the primary roots of desert plants. According to Fahn (1989), these are : (1) a reduced number of cortical layers, thereby shortening the distance between the soil and the stele, and (2) wide casparian strips, which imply grater efficiency of the endodermis.

9.3.2
Types of Anatomical Structures in Desert Plants

Evenari et al. (1971) stated that the basic structure of the transpiring and photosynthesising organs of the desert plants, that is, leaves and, for some plants, stems, reveals five basic types; as usual, there are also transitions between types:

1. Scleromorphous stems and leaves. The term scleromorphous is derived from the Greek words *skleros* (hard) and *morphe* (form). Evenari et al. (1971) stated that they intentionally did not use the more common term "xeromorphic"because it implies a relationship between structure and a dry (xeric) habitat. It is worthy of mention that xeromorphism is not confined to xerophytes, and not all xerophytes exhibit xeromorphic characteristics (Fahn 1989). In this context, one should quote the Maximov's definition of xerophytes (Maximov 1931). He defined them as plants that grow in arid habitats whose transpiration decreases to a minimum under conditions of water deficiency. Generally, we believe that xerophytes can transpire at high rates when they are provided with ample water supply. Nevertheless, they can reduce their transpiration to a minimum when subjected to water deficiency.
   Scleromorphoic plants are characterised by the presence of tissues (sclerenchyma) composed of thick-walled cells that give their organs mechanical strength. They are also characterised by epidermal cells with very thick cuticles and outer walls, sunken stomata, and sometimes by an epidermis consisting of more than one layer of cells. Very often, the cells, especially those carrying out photosynthesis, are smaller than normal. The photosynthesising leafless stems such as those of *Retama raetam* and *Leptadenia pyrotechnica* provide the best examples of scleromorphy. Scleromorphous plants are all built along the same lines, irrespective of the families to which the plants may belong. Examples include *Calligonum comosum* (Polygonaceae); *Deverra tortuosa* (syn. *Pituranthos tortuosus* (Umbelliferae) (Fig. 55), *Lavandula coronopifolia* (Labiatae) (Fig. 56), *Ephedra alte (Ephedraceae)*; *Asparagus stipularis* (Liliaceae); *Ochradenus baccatus* (Polygonaceae); *Zilia spinosa* (Cruciferae), and *Retama reatam* (Leguminosae). This near uniformity of anatomical structure is hardly strange, since most plant stems are constitutionally scleromorphous, irrespective of any photosynthesising functions. Whereas leaves are characterised by great structural variety, stems follow basically similar structural plans. The specific anatomical feature of photosynthesising stems is the presence of green tissues at their periphery. Fahn (1989) described the anatomical structure of the *Retama raetam* stem as follows: "Along the green branches of this plant there are ribs and furrows. The stomata are situated in the furrows in which there

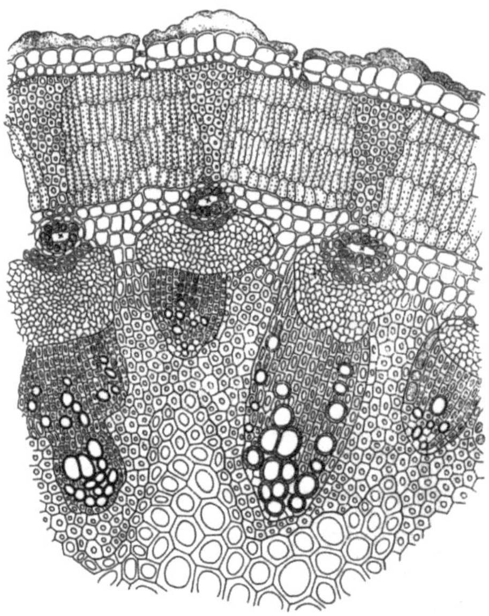

Fig. 55. A transverse section in the stem of *Deverra tortuosa* (syn. *Pithyranthus tortuosus*). (After Volkens 1887)

are also numerous hairs. The central portion of the ribs consists of a sclerenchymatous strand that is accompanied laterally, on the sides facing the furrows, by one or two rows of large colourless parenchyma cells that serve as water-storing cells and usually contain crystals. Between these cells and the epidermis is the photosynthesising tissue, which consists of small and dense parenchyma cells containing chloroplasts and crystals. The epidermal cells have thick outer walls and cuticle".

2. Malacomorphous leaves. This term is derived from the Greek *malakos* (soft) and *morphe* (form). By far the majority of desert plants, both annuals and perennials, summer active and inactive alike, have leaves that do not carry a single scleromorphous feature. Instead, they are characterised by thin cuticles and outer cell walls of the epidermis. Furthermore, the stomata lie in the epidermal plane, have large cells, and possess no special mechanical tissues. The anatomical structure of the malacomorphous leaves of desert plants is generally very different from that of the other plant groups. However the exceptions of the malcomorphous leaves are similar to the majority of leaves of desert plants, including scleromorphous and succulent ones. In many cases, especially among annuals, the leaf tissues consist of more or less round, undifferentiated, isodiametrical cells; an example is *Calendula arvensis* (Compositae). In others, it is made up of palisade cells only, as in *Herniaria hemistemon* (Caryophyllaceae); *Artemisia herba-alba* (Compositae), *and Stachys aegyptiaca* (Labiatae), or of palisades on both sides of the leaf with spongy tissue in the middle, as in *Reseda muricata*. There are all kinds of

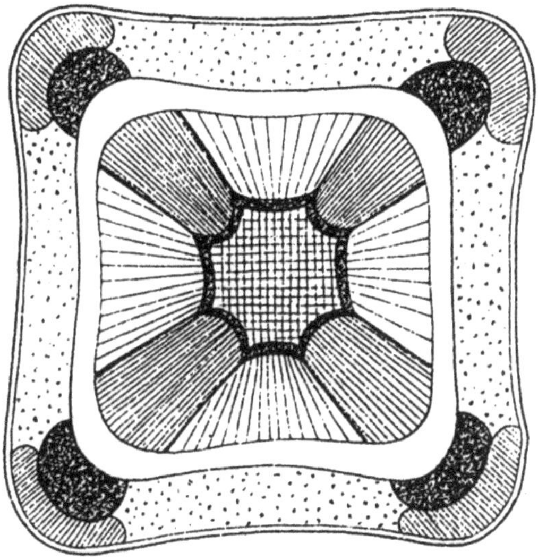

Fig. 56. A transverse section in the stem of *Lavandula coronopifolia*. (After Volkens 1887)

transitional types between those mentioned, but one outstanding feature is common to all: with the exception of the epidermis, the connection between the cells of the leaf tissue is loose; the cells just touch each other and the intercellular spaces are very large.

3. Enchylomorphous (succulent) leaves. The leaflets of different *Zygophyllum* species provide good examples of leaf succulence. These leaflets contain a considerable amount of water stored in the large cells of the centrally located storage tissue. They are surrounded on all sides by much smaller, green photosynthesising cells (Fig. 57). There is no dividing line between the two tissues. In other succulent leaves, such as those of *Mesembryanthemum forskahlei* (syn. *Opophytum forsskalei*) (Fig. 58), the two tissues are sharply separated. A typical succulent leaf has no scleromorphous characters: its cuticle and outer epidermal cell walls are especially thin, and the number of stomata per unit surface is small. The leaf is either cylindrical or flat and very thick. In either case the ratio of external leaf surface to volume is small. Since the great majority of leaves of desert plants have a slight tendency to succulence, this is true to some degree of all leaf types.
4. Leaves with malacomorphous tissue covered by a scleromorphous epidermis. Such leaves seem to be a constant feature of most of the species belonging to the lily family. The leaf tissue is loose and of the same type as the purely malacomorphous leaves of *Allium rothii*. In some cases, as in *Tulipa amplyophyla*, the cells are elongated and parallel to the leaf surface. The epidermis has thick cuticle, outer cell walls and sunken stomata.

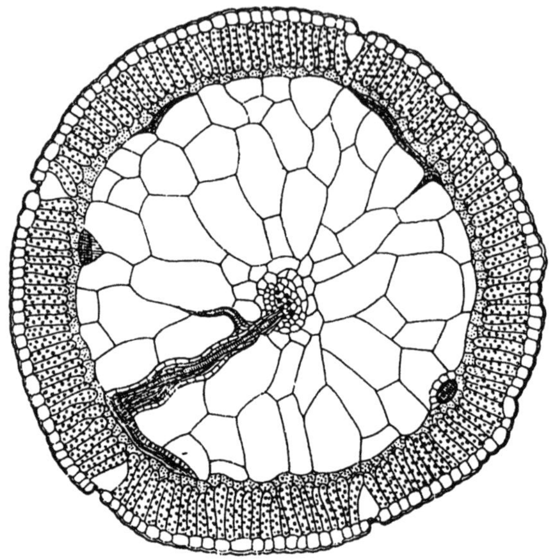

Fig. 57. A transverse section in the leaflet of *Zygophyllum simplex* showing the water storing tissues surrounded by the photosynthetic tissues. (After Volkens 1887)

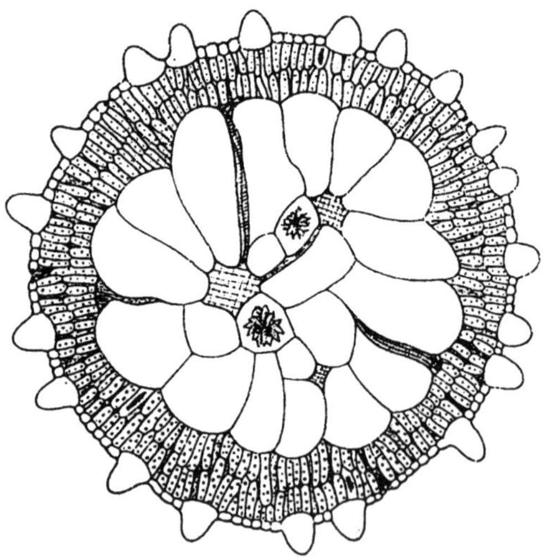

Fig. 58. A transverse section in the leaf of *Mesembryanthemum forsskahlei* showing the water storage cells surrounded by the photosynthetic tissue. (After Volkens 1887)

5. Stems and leaves with an enchylomorphous core and a scleromorphous epidermis. The term enchylomorphous is derived from the Greek *enchylos* (succulent) and *morphe* (form). This arrangement is typical of the photosynthesising stems of many members of the Chenopodiaceae (Fig. 59) and in some cases also of their leaves. The initially succulent petioles of *Capparis spinosa* develop a most scleromorphous epidermis during the summer.

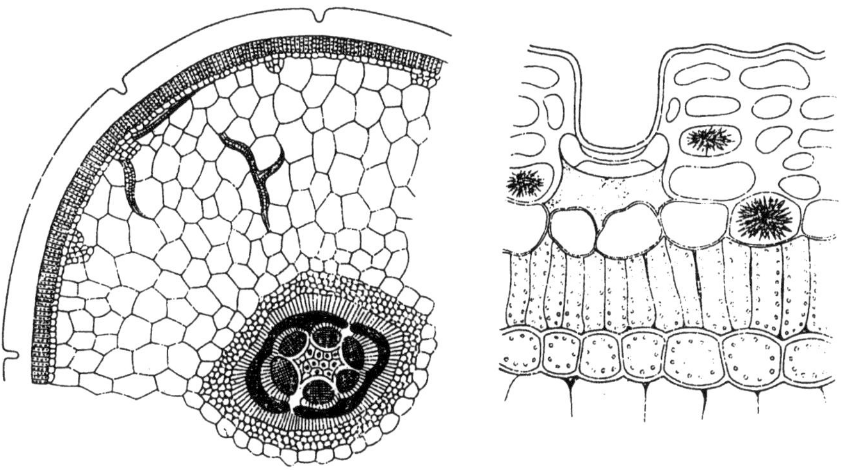

Fig. 59. A transverse section in the stem of *Anabasis articulata* showing the distribution of the photosynthetic tissues. An enlarged part of the section shows the outer layers. (Volkens 1887)

Succulent leaves have a water reserve on which they can draw. The cell walls of their tissue are thin and elastic and can either shrink or expand according to their water content. Despite their water reserve and low rate of transpiration, their faculty for limiting transpiration is low and, contrary to common opinion, they cannot be considered to be especially drought-resistant. This observation does not apply to plants like cacti (not represented in the Middle East flora), where the stem is succulent, contains an enormous amount of water, and is protected by a highly scleromorphous skin. Structures similar to those of cacti may be observed in succulent euphorbias in the southern part of the deserts of the Middle East, such as in Yemen and Oman (Fig. 53).

The succulent stems of desert plants are characterised by a well-developed water-storing tissue in the cortex. This cortex can be shed under dry conditions to reduce the transpiring surface.

9.3.3 Shedding of the Cortex

It is known that in the articulated Chenopodiaceae, such as *Haloxylon* spp. (Fig. 26), *Anabasis* spp. (Fig. 27), and *Arthrocnemum macrostachyum* (Fig. 70), the fleshy photosynthesising cortex is shed from the mature stems in the summer as a result of the formation of a periderm that develops in the phloem parenchyma (Fahn 1963). In some desert plants that do not have a fleshy cortex, such as *Atriplex halimus*, *Zygophyllum dumosum*, and *Fagonia cretica*, the newly formed periderm also develops deep within the stem in the pericycle or phloem parenchyma. This is an adaptive feature.

9.3.4
Formation of Interxylary Cork

In some desert plants, such as *Artemisia* spp. and *Achillea fragrantissima*, interxylary cork rings are produced at the conclusion of each annual increment of wood. Moss (1940) has already pointed out the important adaptive value of interxylary cork tissues. They reduce water loss and restrict the upward passage of water to a narrow zone of functioning secondary xylem (Fahn 1989).

9.3.5
Cell Size

Reduction in cell size has often been noted in plants that have developed under conditions of water stress (Henckel 1964). Cutler et al. (1977) recently restated the case for the importance of small cell size to drought adaptation. They measured osmotic potential (Ns)_/cell, cell size and cell wall thickness in leaves of cotton plants (*Gossypium hirsutum*) grown under water stress and found statistically significant differences in cell size (32 vs 37 µm), in Ns (an increase of about 3 bar) and in cell wall thickness (0.76 vs 0.68 µm) between the stressed plants and the controls. Rather similar observations were made by Wilson et al. (1980) on *Panicum maximum* and *Heteropogon contortus*. Drought-adapted plants tend to have smaller cells with thicker walls than mesic plants. Wilson et al. (1980) stated that they remain unconvinced that the adaptive advantage of small cells results from a decrease in transpiration, which is wholly consequent upon the small size of the cells. The case relating cell size or structure to function is an interesting one, but is still open to speculation, observation and experiment.

CHAPTER 10

Adaptations That Engender Escape from Drought

10.1 Drought Avoidance

Avoidance simply means that growth is confined to periods whenever there is ample water supply or at least sufficient water supply to sustain the life of the plant. To cope with the prolonged annual dry period, which occurs in almost all the deserts of the Middle East, different groups of plants have evolved different survival mechanisms. Evenari et al. (1971) stated that plants are active only as long as the upper soil layer is wet, becoming inactive during the dry season. According to these authors, they can be classified into three groups, namely: winter annuals, plants with bulbs, corms or rhizomes (Geophytes, cryptophytes), and diminutive perennial herbs (nano-phanerophytes, hemicryptophytes).

During the evolution of higher plants, differentiation of vascular tissue and concomitant increases in plant size and complexity led to the development of drought-avoidance mechanisms in species that are adapted to arid regions. These adaptations have the advantage of enabling plants to remain alive during times of restricted water supply. As is now obvious, higher plants are not restricted to this course, despite the intricacies of their structures and physiologies. Most are genetically capable of producing desiccation-tolerant reproductive propagules (seeds).

Avoidance is achieved by the disappearance of the green, transpiring organs. This can be complete disappearance, as in annuals (ephemerals), or disappearance for a definite period during the dry season, as in many geophytes (usually bulbous) and diminutive perennials. However, there are many differences among these groups. This will be explained in the following sections.

10.2 Winter Annuals

In winter annuals, seed production is vital for survival. Because the length of the growing period is restricted by drought, it is necessary for these species to reproduce before conditions become too unfavourable. The majority of winter annuals in the desert show features of species optimally adapted to reproduction in a habitat with a restricted growing season. Growth in the vegetative phase is followed by a phase of almost solely reproductive growth. These plants are called ephemerals because of the brevity of their life cycles.

Annuals are usually herbaceous although their stems may lignify during senescence (e.g. *Anastatica hierochunta*). They have a short life cycle, which may be completed either within 2–3 weeks (*Linaria haelava, Trigonella stellata, Plantago ovata, Schismus barbatus,* etc.) or within a year or more; the individual plant usually dies after seed formation. Annuals normally survive as seeds during the unfavourable dry seasons. Since the survival of annuals, more than that of any other plant life form, depends on the ripening of new seeds, the formation of seeds within a short production period is characteristic. Annual plants are most successful in semi-arid and arid regions, where they may predominate over other plant forms after rainfall, even though perennial woody plants appear to be the dominant plant form during dry periods.

Winter annuals, whose life cycle is restricted to the rainy season, are an especially important component of the desert vegetation during the wet season in rainy years. The percentage of these annuals in relation to the total number of species in any desert is usually higher than 50 % but decreases with increase in rainfall. For example, only 28 % of the plant species are winter annuals in the Mediterranean region of Israel, whereas they represent 59 % in the Negev (Evenari et al. 1971). In Qatar, as Batanouny (1981a) wrote: "The majority of the flowering plants of the flora of Qatar is represented by therophytes. The elimination of such a high number of plants in summer is an important means to minimize the water output from the vegetation. Phanerophytes are represented by six species (almost 2 % of the flora)." In even more extreme deserts, such as certain parts of the Sahara, which may experience complete drought for a number of consecutive years, the percentage of winter annuals is even higher and annual plants may be the only ones to appear after the rare rainfalls. By contrast, shrubs, dwarf shrubs, trees, geophytes, and hemicryptophytes are rarely found in the extreme deserts – an indication that winter annuals are particularly well adapted to desert conditions. Their adaptational mechanisms differ in many respects from those described in other xerophytes. Their roots are mostly shallow (Figs. 14–16) and restricted in extent; their transpiration rate varies only slightly because these plants cannot reduce their transpiring surface to any effective degree; and in the majority of cases they do not tolerate large water deficits. But they do possess the remarkable ability of being able to regulate their total body size according to the moisture conditions of their habitat. When growing in an extremely dry locality *Anastatica hierochuntica*, for instance, is a pigmy, a few millimetres high, with two or three branches, a few small leaves, and one to five fruits. Where soil moisture is plentiful, however, the plants reach 15–20 cm in height, attain a diameter of 25–30 cm, possess tens of branches, and have hundreds of fruits. Variable body size is also a measure of the great variability of developmental speed in winter annuals, since small plants go through their full life cycle in a much shorter time than large ones. This is an important factor in their survival because, in a habitat where water is scarce, the time of optimal conditions is much shorter than in a well-watered environment

Comparing ephemerals from the Sahara with ephemerals from middle Europe, Lemée (1953) found that there is no difference in their morphological features, rates of transpiration or resistance to dehydration. He assumed that their great capacity for water absorption is the main adaptation to the desert climates. However, their adaptational mechanisms are different from those of true xerophytes. The main adaptational trait is the disappearance of these plants before the advent of the dry season.

Other adaptational traits of winter annuals are their mechanisms of germination and seed dispersal that enables the plants' dispersal units either to lie dormant in the ground for long periods as a reserve when conditions are unfavourable for germination, or to germinate rapidly when the environment is favourable. Gutterman (1993) has discussed the adaptation of seed germination in desert plants in one of the books of this series, *Adaptations of Desert Organisms*. For annuals, much more than for perennials, the main danger lies in the enormous variability of environmental conditions between one year and another and also from habitat to habitat. Annuals counteract this by being endowed with an even greater variety of morphological, developmental, and physiological responses than are perennials. These provide an ever-present reserve of viable seeds lying dormant in the soil, ready for all possible climatic eventualities. In this context, one might mention the case of *Reseda decursiva* seeds, which do not germinate if they are collected from the mother plants and kept at laboratory temperature. However, exposure of these seeds to a temperature of 60 °C for 4 h every day over 4 days causes them to germinate with almost 100 % success. This shows that, in many cases, the seeds of annuals need to be exposed to the naturally very hot and dry conditions of the desert. Such conditions help to break the dormancy of these seeds. The life cycles of annuals are closely related to and affected by the prevailing conditions in the desert.

It is informative to consider the growth of ephemerals in a small shallow depression in the desert. The concave shape of the depression causes differences in the thickness of the surface deposits and the amount of water available to the plants in the middle of the depression and on its margins. Individuals of the same species exhibit different growth rates along a line across the radius of the depression. Those individuals growing along the margins receiving a low amount of runoff water and, inhabiting thin deposits, have a low rate of growth and flower earlier than do those in the middle of the depression. This indicates that the extent of drought avoidance differs among seedlings within a locality consisting of different micro-environmental conditions.

In the period extending from late autumn to early summer in many parts of the deserts of the Middle East, ephemeral growth exhibits considerable variation. This variation is closely related to the timing of the rainfall. Early rains, combined with the low temperatures of December, result in the emergence of a particular ephemeral population that flowers in early spring. Late rains, combined with mild temperatures in March, stimulate the appearance of another group of ephemerals that flower in late spring. There is a rhythmic variation in the species that emerge during the same season. These successions of species are obvious in years with evenly distributed rainfall. Temperature conditions control the appearance and flowering of each group of species. The temporal distribution of ephemerals throughout the growing season is a means by which each group of ephemerals fulfils its requirement for certain environmental components at a definite time. If all the naturally occurring seed populations were to germinate at the same time, the severe competition that resulted would lead to the disappearance of some species. Temporal distribution is also important in the water economy of plants growing during the same season. It is also helpful in exploiting the seasonal distribution of pollinating insects.

We should also distinguish between winter ephemerals, which abound within the northern part of the deserts of the Middle East, and summer ephemerals that predominate in the southern part of the region when rain falls during the summer months.

Despite the strict division between annuals and perennials, certain plant species have been recorded as "biennials" by some investigators or as either annuals or perennials. Species such as *Zygophyllum simplex*, *Arnebia hispidissima* sand *Sclerocephalus arabicus* have been observed to persist until the following growing season.

10.2.1 Potential Annuals

In his article on potential annuals in the Egyptian desert, Haines (1951) wrote: "But in a good season, such as the winter of 1949–1950 (when there were 13.7 mm of rain in November and December), many perennials succeed in flowering and setting good fruit in their first season, so that even if they are killed by the drought of the following summer they have completed a full life cycle as potential annuals, and this ability may be an important biological factor in their survival." Examples given by Haines include: *Diplotaxis harra* (Forssk.) Boiss., *Cleome africana* (syn. *C arabica* L.), *Caylusea hexagyna* (Forssk.) M.L. Green (syn. *C. canescens*), *Farsetia aegyptia* Turra and *Zilla spinosa* (L.) Prantl.

10.3 Geophytes

The bulbs, corms or rhizomes of geophytes buried in the soil remain dormant during the dry season, well protected against drying out and overheating by layers of dead tissue or cork. While in this inactive state, they need very little water, if any, to remain alive and so evade the drought conditions of the environment. The dormant bulbs of *Poa sinaica*, for example, can be heated to 80 °C without suffering any damage. However, shortly after the first rains and as soon as the soil in the vicinity is wet, they form roots and rootlets with incredible speed. Less than 12 h after the first rain has wetted the upper 2 cm of soil the first rootlets appear on *Carex pachystilis*, and a few hours later their dense network enmeshes the upper soil layer. Examples of bulbous geophytes growing in the deserts of the Middle East include *Dipcadi erythreum*, and *Allium desertorum*; of the geophytes with corms, *Colchicum rotchii* and *Arisarum vulgare* are typical.

Scilla hanburyi is a geophyte from the desert of the Negev with two active periods during each year (Evenari et al. 1971). The first growing period occurs at the end of the dry summer, when the plant puts out its slender flowering stalks carrying small, pale violet-rose flowers. No leaves appear at this time and, after seed formation, the plant returns to dormancy. The second active period begins after the first rains and only leaves are formed. The plant does not flower and the leaves dry out and die in February–March. At least two other Negev geophytes behave similarly, *Urginea undulata* and *Pancratium sickenbergeri*. Both species grow widely in the deserts of the Middle East. The same behaviour has been observed in the case of *Urginea maritima* (Batanouny and Khalifa 1970), which has very big bulbs (Fig. 60). It is a source of scillarin-a and scillarin-b, two cardiac glycosides. The extract of the plant is an ingredient of expectorants and the bulbs are used in folk medicine (Batanouny 1999).

Fig. 60. *Urginea maritima* in winter, the wet season. Note the green broad leaves and the large bulbs

10.4 Diminutive Perennials

The behaviour of the hemicryptophytes is best demonstrated by *Erodium crassifolium* Willd (syn. *E. hirtum* L'He'r) (Evenari et al. 1971). This plant is a small dwarf herb that is inactive in summer. It has comparatively large dissected leaves and large violet-lilac flowers, which in their profusion contribute much to the blazing colours of the slopes whenever the desert blooms. The woody stems are very small and grow close to the ground. Their annual permanent increment in length and thickness is insignificant and the stem of a plant tens of years old is still very small. These stems carry renewal buds, which develop leaf- and flower-carrying shoots after the first rains and which are shed at the beginning of the dry season. Only the leafless woody stems remain to survive the summer in a dormant state. The main roots of *Erodium crassifolium* penetrate more than 40–50 cm into the soil, exploiting a relatively large volume of soil for water. Some of the roots develop fleshy tubers containing a high percentage of water and sugar. These are a favourite food of the Bedouin. Evenari et al. (1971) gave many other examples of diminutive perennials such as *Erodium glaucophyllum* (L.) L'Hér, *Herniaria hemistemon* J.Gay, and *Reseda muricata* C.Presl, which belong to this same group of diminutive dwarf herbs. The water economy of *Erodium hirtum* and *Reseda muricata* is typical of all hemicryptophytes and, with one most important exception, is also similar to that of the summer-active *Haplophyllum tuberculatum* and *Heliotropium rotundifolium*. The exception lies in the fact that the transpiration rate of the leaves during the rainy season is high, but once the soil starts to dry out the transpiring surface is much reduced. The plants can tolerate a relatively high saturation deficit before they gradually lose their leaves and become dormant. When the hemicryptophytes are no longer capable of supplying their daily increasing water deficits, they shed all their shoots. The duration of their active period is a function of the varying moisture conditions of their habitat. In relatively good years it may last for 4 months, whereas in dry years it may be restricted to 5 or 6 weeks. In a restricted active period very few leaves are formed, and flowering is often suppressed.

CHAPTER 11

Adaptation to Desiccation

11.1 Resurrection Plants

Desiccation tolerance is a feature of plants capable of experiencing protoplasmic dehydration without permanent injury. Except for the seed stage, this latter mechanism is poorly developed in the majority of higher plants. Losses greater than 40–90 % of normal water content (Levitt 1972), or equilibration with relative humidities lower than 91–98 % (Levitt et al. 1960) are lethal for most higher plants. By contrast, many lower plants (and some angiosperms) are remarkable in the ability of their vegetative tissues to withstand extreme drought and desiccation. Known also as poikilohydrous plants (Walter 1955b), poikiloxerophytes (Henckel and Pronina 1968), or resurrection plants, these desiccation-tolerant plants are loosely defined by their unique ability to revive from an air-dry condition (the air having a low relative humidity). Plants that succumb to desiccation are called desiccation-sensitive, desiccation-intolerant or homoiohydrous plants (Walter 1955b).

A wide variety of plants appear to be able to survive extremely low water potentials for prolonged periods. In addition to thallophytes, such as many algae, lichens, and mosses, about 83 species of angiosperms and 60–70 species of ferns and fern allies can be dried to equilibrium with water vapour at 10 % relative humidity or less (<–3,000 bar) for many months. After drying they are still viable and resume metabolism within a few hours or days after rehydration (Gaff and Hallam 1974; Gaff 1980; Bewley and Krochko 1982). Most of these desiccation-tolerant angiosperms are native to southern Africa, with a few from other regions but none from North America or the Middle East deserts.

11.2
Desert Poikilohydrous Thallophytes

Knowledge about the life and adaptations of thallophytes in the deserts of the Middle East is still far from complete. The adaptations of algae to arid conditions have been studied by Berner (1974) in the Negev, while the adaptations of lichens to desert conditions were investigated by Lange (1965b, 1969a,b), Lange et al. (1969, 1970a,b, 1975), Lange and Evenari (1971) and Kappen and Lange (1972). The distribution and sociology of terrestrial bryophyte communities were studied by Frey et al. (1990; Frey and Kürschner 1992) in the Judean desert. Frey and Kürschner (1991a–c) investigated the adaptation and life strategies of bryophytes in arid and semi-arid regions. These lower plants are distinguished from flowering plants by being poikilohydrous (Evenari et al. 1975). They obtain their water requirements from the surrounding atmosphere through their surfaces and are able to withstand extreme and repeated dehydration without being damaged. Also, they are able to absorb water from highly humid air and to photosynthesise after a short time. They therefore depend on dew and air humidity, which are more stable than the erratic rainfall under desert conditions.

Poikilohydric plants can occupy habitats that are unavailable to intolerant plants because, relative to potential carbon gain, they can benefit from: (1) extreme desiccation (and sometimes temperature) tolerance; (2) utilisation of dew or water vapour for hydration (some species only);and (3) rapid emergence from the quiescent state, including photosynthetic recovery with the availability of water.

Nonetheless, these benefits must be considered against the potential carbon costs associated with: (1) stored energy and/or time costs for physiological reactivation upon rehydration; (2) physiological deterioration with increasing time in the dry state, (3) low photosynthetic rates, and (4) decline in membrane integrity permitting carbohydrate and nutrient losses during rehydration with liquid water.

Different species have developed different ecological, morphological, and (presumably) physiological adaptations to ultimately accomplish a net gain in their photosynthetic productivity (Bewley and Krochko 1982).

Apart from bacteria, lichens are the most successful pioneers in extreme environments: members of this group can tolerate very hot, very cold, and very dry conditions and some species are even found on permanently inundated habitats (Lange and Ziegler 1986).

There do not appear to be any common morphological features that distinguish poikilohydrous plants from intolerant species. Tolerance of desiccation can be attributed largely to properties of the protoplasm. Tolerant plants do not undergo severe water deficit without perturbation to their metabolic and cellular integrity, but critical features of desiccation-tolerance may be the ability to limit damage caused by desiccation to a repairable level and to repair damage quickly upon subsequent rehydration. The integrity of the majority of metabolic systems (or at least of their components) is probably retained in the dry state, and these systems are quickly reactivated upon rehydration. This is probably also an important feature, since hydration of many poikilohydrous plants is a transient phenomenon. Obviously, any plant in any environment can only survive if it is able to achieve a net gain in its carbon status, and the adaptations of poikilohydrous plants should be viewed in relation to this fact (Eickmeier 1979).

When poikilohydrous plants are dry, they become dormant and their metabolic processes are reduced to a minimum. In this state their reactivity to the environment is much reduced. According to Evenari et al. (1971), they are physiologically and ecologically quasi-isolated from their surroundings and therefore more resistant to drought and high temperature. Because of their activation and inactivation cycles, desert thallophytes have limited periods during which the production of dry matter takes place. Therefore their rate of growth is extremely slow.

11.2.1 Algae

Orshan (1986) stated that the green algae *Protococcus grevillii* and *Protosiphon cumoreii* and the blue-green *Nostoc* sp. and *Microcoleus* sp. form an algal crust on loess soils so that even after a light shower the hills and plains of the Negev upland turns greenish, unveiling the surprising fact that the desert is in fact an algal ocean.

According to Berner (1974), desert algae can absorb water from the air when the relative humidity exceeds 70 %. They become photosynthetically active within 15–20 min after being wetted. The same author, as quoted from Orshan (1986), estimated the standing algal biomass to be 0.02 t/ha.

The lower sides of semi-translucent flints in the desert represent a habitat for particular algae. They enjoy the relatively high humidity conserved below the flint and receive light through the flint window. These hypolithic algae can photosynthesise at lower light intensities than algae in the open desert. Slight showers moisten the thin soil film below the flint. Such moisture is protected from rapid evaporation and represents a reasonable water supply to the algae below the flint, which make use of the humidity of the air.

11.2.2 Lichens

The number of studies on the lichens of the deserts of the Middle East is limited. Among these one might mention: Nylander (1864), Müller (1884,

1891), Kneucker (1926), Galun and Reichert (1960), Galun (1970) and Galun and Garty (1972). In the last mentioned, the authors recorded 51 lichen species (belonging to 11 families and 22 genera) from northern and central Sinai. The adaptations of desert lichens to drought and extreme temperatures were investigated by Lange (1965b, 1969a,b), Lange et al. (1969, 1970a,b, 1975), Lange and Evenari (1971) and Kappen and Lange (1972). The special physiological adaptations that enable the lichens to live under hostile desert conditions will be summarised from these investigations.

The responses to drying of the whole lichen thallus are a combination of those of the algal (phycobiont) and of the fungal (mycobiont) component. Desiccation-induced changes in photosynthesis, for example, reflect the metabolic state of the algal component which, although it is only 3–10 % of the thallus by weight, is the fixer of carbon that is subsequently transported as carbohydrate to the more massive fungal component. The latter is the major site of respiration. Several reviews have been written about the effects of environmental stresses on lichen survival and carbon assimilation (Ferrar 1973, 1976; Kappen 1973; Lange et al. 1975; Smith 1975; Harris 1976) and the reader should consult these for details that are not presented here.

Recovery from desiccation has been studied under both field and laboratory conditions. In their natural habitat some desert lichens are rarely re-wetted by rainfall, but instead resume their metabolism in atmospheres of high relative humidity and during early morning dew formation. Dry *Evernia* and *Ramalina* species can absorb sufficient water from moist air to resume CO_2 exchange; in fact, at relative humidities above 80 % respiration and photosynthesis increase nonlinearly with increasing moisture content of the ambient air (Bertsch 1966; Lange 1969a). These lichens can achieve at least 90 % of their maximum activity in water-saturated air, and yet under these conditions (when care is taken to prevent water from condensing into droplets) the water content of *Ramalina maciformis*, for example, only reaches 35 %. The compensation point of this species is reached at 80 % relative humidity (10,000 lx, 10 °C), corresponding to a water content of about 20 %, and a water potential close to −300 bar. Fixation of radioactive CO_2 by photosynthesis in *Lepraria membranacea* has been observed at water potentials as extreme as −450 bar (Brock 1975).

If *R. maciformis* is allowed to absorb water from moist air (to a water content of 37 %), there is an increase of only 20 % in CO_2 assimilation upon subsequent wetting of the thallus with liquid water (to 60 % water content); however, respiration in darkness more than doubles (Lange 1969a). The reasons for this are not known, but these observations are of relevance to the environmental conditions to which this lichen is subjected in its natural habitat in the desert. The intervals between winter rainfalls may exceed 12 months, but this and other lichens are often moistened at night by humid air and by early morning dew. Before dawn the water content of the thallus *R. maciformis* may reach 31 %, but the night temperatures are too low to allow for much respiration. As dawn breaks and the light intensity increases a net assimilation of CO_2 occurs, but photosynthesis then declines within 2–3 h as the thallus loses water to the now drier, and warmer, air. Respiration also resumes, although less rapidly, but by the time optimal temperature conditions have been reached the thallus has dried out again (Lange et al. 1970a). It has been calculated that on a 24 h basis the CO_2 yield of dry weight by assimilation is 1.32 and loss by respiration is 0.78, This gives a net yield of 0.54 mg/g dry weight, or an equivalent of 146 μg of organic carbon fixed (summarised by Kappen et al. 1975; Lange et al. 1975). Further calculations, based on the fact that *R. maciformis* is normally subjected to 198 dew events per year in the central Negev, suggest a yearly increment of thallus growth of 8.4 %. Not all desert-inhabiting lichens resume photosynthesis in response to increasing air humidities (Gannutz 1969). *Chondropsis semiviridis*, a native of the Australian deserts, probably requires liquid water (dew or rain) before resuming photosynthesis (Rogers 1971).

11.2.2.1
Water Uptake by Lichens

The lichens that absorb dew are capable, like algae, of absorbing water from air with a relative humidity greater than 70 % (Orshan 1986). In the early hours of the morning, when daylight begins to break, they begin assimilating; but they dry up rapidly after sunrise and assimilation ceases until the next morning when they are wetted again (Lange et al. 1970a,b).

Under arid conditions, rates of gas exchange may vary substantially in different plant species, due to differences in structure and metabolism.

Lichens are exclusively dependent on water uptake from their surroundings by the whole thallus surface. Their internal water potential changes with the change in the ambient water potential. The sources of water received by the thallus of desert lichens include the meager precipitation (rainfall and dew) and the water vapour of the air after it has reached a high vapour pressure even though it is still below saturation. Reactivation of desiccated lichens occurs after rain-showers (a limited number of days every year), after receiving nightly dew fall, and/or after exposure to humid air.

The water relationships of lichens have provided an active subject for research, as shown in recent reviews (Blum 1973; Ferrar 1973; Harris 1976). Most of the data on the subject have, however, concentrated on the ecological and physiological responses to changes in thallus water content. The biophysical and biochemical consequences of water uptake in lichens are very poorly understood, particularly with respect to the absorption of water vapour.

Water uptake, as well as its loss, are rapid and purely physical processes. There are no strong data to suggest any active metabolic control of the uptake process (see Blum 1973). Nevertheless, the morphological and anatomical structure of lichen thalli may be very significant in determining these rates. In addition to uptake from precipitation, runoff, and dew, the majority of lichens are able to hydrate beyond their moisture compensation point in humid atmospheres and this property sets them aside from most bryophytes and vascular plants.

Uptake of liquid water by lichens is a rapid process that takes place over the entire thallus surface. For most species, full saturation is reached in 1–4 min or less (Smith 1962; Blum 1973). This rapid uptake is thought to occur both by absorption of water by the thick-walled fungal hyphae and by capillary movement external to the hyphae. At a stage nearing saturation, the capillary water content of most foliose and fruticose lichens is 100–300 % of the dry weight of the thallus. Limited data for crustose lichens suggest similar values (Ried 1960; Blum 1973). No significant differences occur in the water contents of fruticose and foliose taxa. However, the gelatinous lichens of the Collemataceae have structural characteristics that allow for their greater hydration.

Differential water storage within lichen thalli is clearly induced by wetting and drying cycles. As the fungal hyphae in the upper cortex swell with the uptake of water, a stage is reached at which gas exchange with the algal layer is restricted. Rates of photosynthetic assimilation of most lichen increase directly with thallus moisture content up to this stage, but frequently decline as the moisture content approaches saturation. During drying, water is lost rapidly from the outer cortex, thereby decreasing resistance to gas exchange. The result is a hysteresis effect with significantly different moisture contents for maximal rates of photosynthetic assimilation in wetting and drying cycles.

Moisture uptake from water vapour and condensation dew is extremely important to many lichen species as we have noted. The ecological and physiological significance of these sources of water has been documented in considerable detail for lichens in the Negev Desert. Lange et al. (1975) summarised these studies. Dew fall in the Negev is frequently sufficient to engender hydration in the early morning, allowing for several hours of active photosynthesis. Over the course of a year, the moisture input, principally from dew, can account for net production rates of 8–15 % of the original biomass of the thallus (Kappen et al. 1975; Lange et al. 1975). This is a level equal to or greater than that of vascular plants in the same region.

Under conditions of high relative humidity, lichen thalli slowly absorb moisture until they reach a an internal vapour pressure that is in equilibrium with the atmosphere. Most lichens achieve a water content of 40–50 % in a saturated atmosphere, but a number of days are required for this stage to be reached. Exposure for several hours to relative humidities above 90 % is usually sufficient to bring thallus moisture contents above the photosynthetic compensation point. It is not surprising, therefore, to observe that areas of the world with consistent patterns of high humidity but low precipitation support rich lichen floras. This diversity and the associated ecological importance of lichens are particularly evident in desert fog zones (Rundel 1978).

It has been found that desiccated desert lichens are capable of obtaining enough water to become physiologically active solely from air with a high water vapour pressure. Reactivation of lichens is possible even at water potentials much lower than that of saturated air. Experiments have shown that respiration is measurable only when water potential equals that of air at 75 % relative humidity. At 80 % relative humidity, which corresponds to a water potential of about -290 atm, however, lichens reach this moisture compensation point. Net photosynthesis at 90 % relative humidity amounts to one fourth of the maximum attainable at 100 % relative humidity. The fact that photosynthetic activity occurs after moisture uptake without any liquid water being available and that productivity begins at a water potential as low as −290 atm illustrates the efficiency of lichen physiology.

Periods of morning activation, either by moistening with dew or by water vapour uptake are terminated in the desert by high temperature and strong desiccation brought about by extreme solar radiation. Desert lichens have developed an unusual tolerance of these situations and do not suffer any damage. They are capable of surviving long periods almost completely hydrated.

11.2.2.2
Lichens and Air Temperature

In addition to their high resistance to drought, desert lichens are extremely insensitive to heat while desiccated. High temperatures are not a constant feature of deserts. On the contrary, the soil surface during winter may fall to temperatures far below 0 °C which is injurious to many higher plants. Experiments show, however, that desert lichens are affected neither in their vitality nor in their metabolic activity by being subjected to very low temperatures.

It is clear how well desert lichens are adapted to the extreme climatic conditions of their environment. Moistening after long periods of drought and extreme heat, as well as after periods of frost in winter can induce photosynthetic activity almost immediately.

The special adaptations of desert lichens to high temperatures, however, exist only under conditions of dehydration. When moistened and metabolically active, lichens show pronounced sensitivity to high temperatures and less adaptation to relatively low temperatures.

The optimal temperatures, however, do not exceed 20 °C. The optimal range is 10–20 °C, which appears at first to be very low for an organism whose natural habitat is a hot desert. However, periods of actual metabolic activity by these lichens are restricted to the cool morning hours just following sunrise, after they have been moistened by dew or by high atmospheric humidity. During that time, the temperature of the thallus does not reach far above 25 °C in summer, and during winter it seldom reaches 10 °C during the same period of the day. This temperature range is optimal for photosynthetic productivity of desert lichens, especially when solar radiation is low.

11.2.3 Bryophytes

Stomata are absent from the gametophytes of bryophytes and gaseous exchange occurs freely over the entire hydrated cell surface. The ability to photosynthesise and respire declines with desiccation of the gametophyte. Reintroduction of water enables these activities to resume. In their natural environments most desiccation-tolerant bryophytes remain dry for several weeks or months between rainfalls, although not all species need to take up liquid water to resume metabolism (Bewley and Krochko 1982). Some mosses recommence photosynthesis after absorbing water vapour from a moist atmosphere (Lange 1969b). A dry period of several days to several weeks probably has little effect on the metabolic recovery of tolerant mosses. It is interesting that the chlorophyll content of desiccation-tolerant species is largely unaffected by drying and dehydration (Bewley et al. 1974; Gupta 1978). By contrast, considerable chlorophyll loss occurs from intolerant species (Gupta 1978).

Terrestrial bryophyte species found along a transect across the Judean Desert from the Dead Sea to Jerusalem are grouped by Frey and Kürschner (1991a) under seven life strategies. These include colonists with generative reproduction, "generative" colonists; colonists with vegetative reproduction, "vegetative" colonists; colonists with bastion innovation, "innovative" colonists; perennial colonists; annual and short-lived and/or perennial shuttle species; and geophytes. Within this spectrum the authors recognised distinct life strategy adaptations in relation to climatic and edaphic conditions. Under extremely arid conditions and on unstable soils rich in salt, taxa with innovative colonist and short-lived and/or perennial shuttle life strategies seem to have decisive advantages. Other strategies have been observed in semi-arid and sub-humid Mediterranean zones. In their study on the morphological and anatomical adaptations of terrestrial bryophytes, Frey and Kürschner (1991b) concluded that the following characteristics could be considered as adaptations to extreme desert conditions: (1) the ability to roll in the thallus, (2) the formation of buds and glass hairs which serve as optimal protective structures against light and temperature, and (3) sunken capsules and short peristomes.

It has been hypothesised that water held in the thick cell walls of xerophytic bryophytes may act as a buffering system for their internal water content during periods of water stress (Hébant 1977; Rundel 1982).

11.2.4 Ferns

The relationship of leaf trichomes to water uptake has been studied in *Asplenium ceterach* (syn. *Ceterach officinarum*), a poikilohydric fern from the Mediterranean region and the Middle East (Roschall 1938; Oppenheimer and Halevy 1962). On rehydration, respiration of this fern resumes at a level higher than in undesiccated controls and requires several days to return to control levels.

The poikilohydrous spikemoss *Selaginella lepidophylla*, the so-called resurrection plant of the southwestern United States, is well known for its rapid foliar uptake of liquid water (Rundel 1982). Desiccation of this fern results in a decline in RUBP carboxylase activity by about 40 %, but recovery of activity occurs within the first 24 h of rehydration – possibly by de novo synthesis of the enzyme (Eickmeier 1979). Thus recovery of photosynthesis after desiccation may be facilitated in this species by a high level of conserved RUBP carboxylase and the ability to quickly restore enzyme levels to normal (Bewley and Krochko 1982).

CHAPTER 12

Adaptation of Plants to Saline Conditions in Arid Regions

12.1 Saline Regions and Halophytes

Saline regions are widespread throughout the deserts of the Middle East. In addition to other factors, this is mainly related to the geographical location of the region. It is bordered on several sides by large bodies of saline water, including the Mediterranean Sea, the Red Sea, the Indian Ocean, the Arabian Sea, the Gulf of Oman and the Arabian (Persian) Gulf. The presence of saline water bodies along the shores of the countries in the region creates vast areas of coastal salt marshes. In addition, there are extensive areas occupied by inland saline tracts. The area of salt-affected land is increasing in the region as a result of anthropogenic factors. It is significant that the Arabs, from time immemorial, have used the Arabic term *sabkha* (pl. *sebakh)* for these saline regions. They described them as being salty and more or less waterlogged in some cases. Only a few plants can grow under the conditions prevailing in the *sebakh.* Though Pallas (Schrader 1809) gave the name "halophyte" to such plants early in the nineteenth century, more than 15 centuries ago, the Arabs gave them the name *hamdh*, which denotes plants with salts in their bodies and growing in *sebakh*. Arabic lexicographers give various specifications for this group of plants, including: salty plants, plants with a sour taste, succulent plants that ooze water when punctured, plants with a disagreeable smell, plants used for washing clothes, utensils and the hands, plants that do not flower in spring but flourish during summer. These characteristics coincide with those assigned to halophytes as defined by modern scientists. It is even more interesting to know that the Arabs give examples of these *hamdh* plants as follows: *rimth (Haloxylon* and *Hammada* spp.), *girm* or *shurah (Avicennia marina*, Fig. 61), *soweid (Suaeda* sp.), *shnan* (*Seidlitzia rosmarinus,* Fig. 62) *ikrish (Aeluropus lagopoides*, Fig. 63*)*, etc. The Arabs differentiate between *hamdh* plants and the *kholla*, which are the sweet plants, i. e. they do not contain much salt compared to *hamdh* plants, which are loaded with salt. Evidently, one can easily correlate these two terms – *hamdh* and *kholla* – with halophytes and glycophytes, respectively.

Fig. 61. *Avicennia marina* (Arabic: *Girm* and *Shourah*). Note the pneumatophores. The Arabian Gulf

Fig. 62. A branch of *Seidlitzia rosmarinus* (Arabic: *Shnan*) showing succulent leaves

Fig. 63. *Aeluropus lagopoides* (Arabic: *Ikrish*) growing in the coastal saline areas along the Arabian Gulf

12.2 Adaptations and Mechanisms

Saline regions are characterised mainly by a high content of NaC1 and a high osmotic potential. High concentrations of sulphates, carbonates, and bicarbonates, and high soil pH, as well as poor soil structure and disturbed aeration may also occur in many saline places. In the Middle East, aridity aggravates the problems created by salinity, and both have an adverse effect on the nutrition, growth, and reproduction of most plants. Halophytes in arid regions, though sharing some characteristics with those in humid places, have acquired various adaptive mechanisms during the course of their evolution. The ability of a plant to survive, grow and reproduce, or even survive without net growth means that it is endowed with particular adaptations that enable it to resist and/or tolerate salinity. Adaptations appear and are incorporated into the genotype, along with those resulting from pressures imposed by other environmental conditions. Such plants spend their lives torn between Scylla and Charybdis! The harmony between the various mechanisms and adaptations is precarious. Exclusion of particular ions and reduction of their absolute or relative concentrations in cells and tissues are undertaken by a variety of mechanisms. Osmotic adjustment and the accumulation of salts is yet another adaptive mechanism. Under arid conditions, plants need to be endowed with mechanisms that enable them to maintain their water balance positive, at least for certain periods of the year. The delicate equilibrium between these various mechanisms can be observed in the life of the plants. A classification of such mechanisms is necessary in order to facilitate our understanding of them and to evaluate the role of each. Numerous investigators (e.g. Arnold 1955; Waisel 1972; Flowers 1975; Breckle 1986; Batanouny 1994) have classified the mechanisms and adaptations under discussion. The main categories are shown in Table 6.

Halophytes growing in a wide range of habitats exhibit adaptations and degrees of acclimation that may differ in nature and/or efficacy. There is kind of plasticity in the responses of these plants, which varies according to the severity of any particular condition. Each species has its own array of adaptive mechanisms. Although halophytes have attracted the attention of scientists for many years, our knowledge of their adaptations is still limited. An attempt is made here to summarise available knowledge regarding the adaptations and characteristics that enable halophytes to live under the adverse conditions of salinity in the arid zone of the Middle East.

Table 6. Adaptive mechanisms of plants to cope with salinity stress. (After Batanouny 1992b)

Avoidance mechanisms
Growth only during seasons of favourable low salinity; such plants have short life cycles
Growth only on favourable sites with low salinity
Limitation of root growth and absorption to distinct soil horizons
Evasion mechanisms
Selectivity against Na^+ and Cl^-
Separation of salts from the assimilating tissue
Compartmentalization of salt within plant organs, tissues and cells
Retranslocation of salts to roots and its excretion by the roots
Leaching of salt out of the shoot
Shedding of salt-loaded tissues and organs
Secretion of salts by salt glands or bladders
Production of adventitious roots and mound building
Tolerance mechanisms
Increasing the salt resistance of tissues, cell and organelles
Production of protectants
Osmotic adjustment
Ion-specific adaptation of the cytoplasm
Increase of halo-succulence; leaf and stem succulence

12.3 Salt Excretion

12.3.1 Salt Glands

Salt excretion is one of the best-understood adaptive mechanisms enabling halophytes to regulate the salt content of their leaves and to cope with the salinity of their environment. This function is performed by specialised glands; these were discovered and described in various species during the late nineteenth century by several investigators (De Bary 1877; Volkens 1884; Marloth 1887). The studies of Schtscherback (1910), Ruhland (1915), De Fraine (1916) and others were followed more recently by Arisz et al. (1955), Ziegler and Lüttge (1966), Lüttge (1971), Batanouny (1973b), Batanouny and Abu Sitta (1977), Batanouny et al. 1992) and many others. In addition to graminaceous species with salt glands, numerous dicotyledonous plant species that possess glands grow in the deserts of the Middle East. They include *Avicennia marina*, *Cressa cretica*, *Frankenia* spp., *Limonium* spp., *Limoniastrum monopetalum*, *Rhizophora mucronata*, *Ceriops* sp., *Bruguiera* sp., *Tamarix* spp. and *Reaumuria* spp.

12.3.1.1 Structure of Salt Glands

The structure of salt glands varies greatly in different plant species but is usually similar in plants within the same family. The simplest type of gland is that found in the Gramineae. Glands are present in longitudinal rows parallel to the veins on both sides of the leaves. Each gland consists of two cells: a basal cell and a cap cell. The basal cell collects salt while the upper cell excretes it. The cells contain dense cytoplasm and prominent nuclei, but lack central vacuoles. The glands are characterised by cutinised and suberised cell walls. Multicellular glands comprising 16 cells are found in the Plumbaginaceae. Some glands excrete carbonates in addition to NaCl, hence they are termed chalk glands. The structure of the chalk (salt) glands of *L. monopetalum* was incorrectly described by Volkens (1884) as consisting of eight cells surrounded by four subsidiary *Nebenzellen*. Batanouny (1973b), however, found that the glands are formed of 12 cells surrounded by four *Nebenzellen*. The 12 secretory cells are arranged in groups of four cells, separated by thin walls at right angles to each other and perpendicular to the surface (Fig. 64). The four subsidiary cells surround the secretory cells. A pit is situated at the top of the gland surrounded by a transverse opening on the leaf surface. The 12 glandular cells have extremely thin walls, except, to a slight extent, those adjacent to their bases. They lack central vacuoles but have a granular cytoplasm and large nuclei. The outer walls of the subsidiary

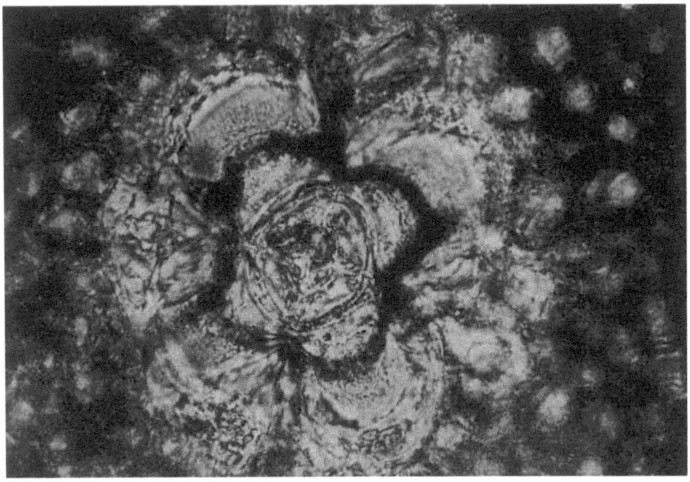

Fig. 64. Surface view of the salt (chalk) gland of *Limoniastrum monopetalum*. (After Batanouny 1973b)

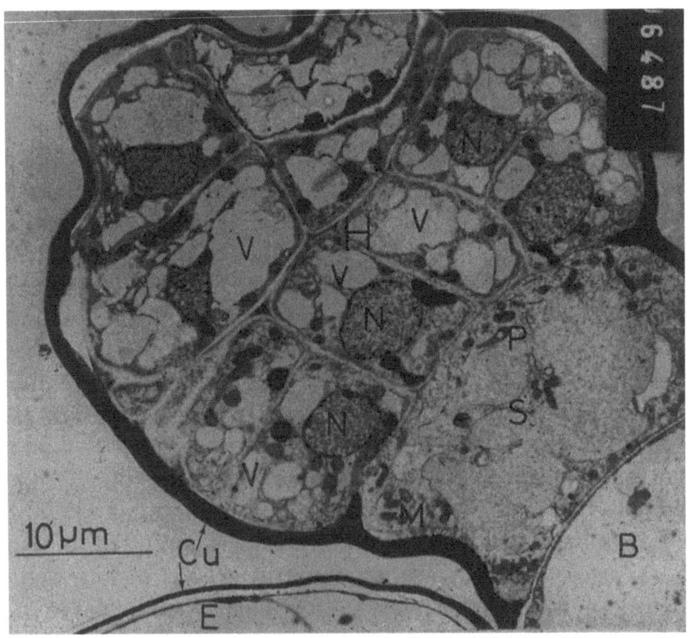

Fig. 65. Electron micrograph of a cross section of the glandular trichome of *Cressa cretica*. The thick cuticle (*Cu*) extends around the trichome and continues with the outer one of the epidermal cells. The anticlinal walls of the stalk cell (*S*) are cutinised. The head cells (*H*) have large nuclei (*N*), numerous mitochondria (*M*), few plastids (*P*) and numerous vacuoles (*V*). Note the basal cell (*B*) and part of the epidermal cell (*E*). (After Fahmy 1986)

Fig. 66. Longitudinal section in the salt gland of *Limonium axillare*. (After Batanouny 1986a)

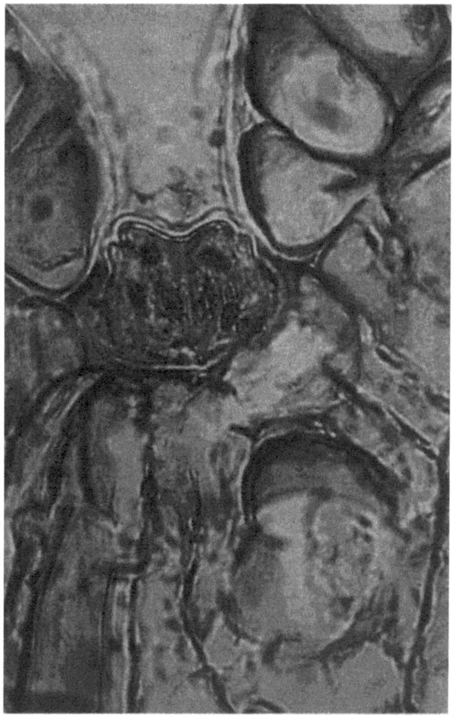

cells are strongly cutinised. A thick cuticle covers the upper sides of the gland and the neighbouring epidermal cells. A rigid structure enclosed by cutinised walls is formed below the leaf epidermis in which the gland is firmly inserted (Batanouny and Abu Sitta 1977). Multicellular salt glands have also been recorded in *Cressa cretica* (Fig. 65; Fahmy 1986) and in *Limonium axillare* (Fig. 66; Batanouny 1986a). The ultrastructure of the salt glands of *Limonium raddianum* (syn. *L. delicatulum*) was studied by Batanouny et al. (1990; Fig. 67). As regards the number and location of the salt glands, the information provided by Batanouny et al. (1977, 1990) is shown in Table 7.

Table 7. Number of salt glands on the upper and lower epidermis of two halophytes

Species	Upper epidermis	Lower epidermis	Reference
Limoniastrum monopetalum	1955–1979	2315–2406	Batanouny and Abu Sitta (1977)
Limonium raddianum	2022	2930	Batanouny et al (1990)

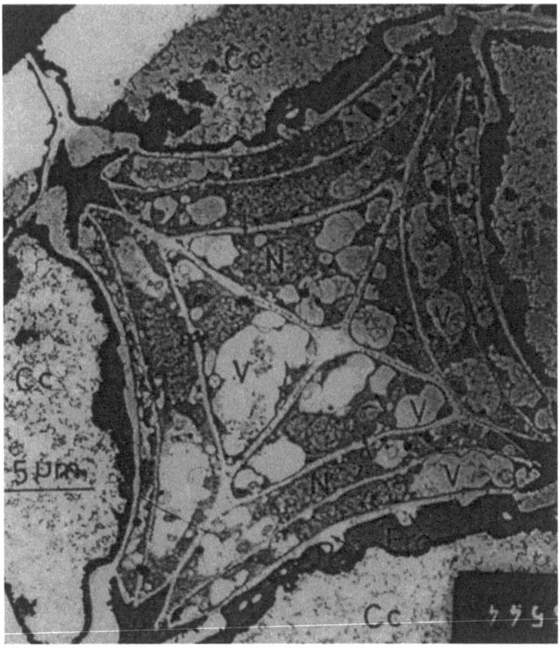

Fig. 67. Transmission electron micrograph of the salt gland of *Limonium raddiana* (syn. *L. delicatulum*) showing the arrangement of the gland cells in three rings: each consists of four cells. *Cc* Collecting cells; *N* nucleus; *pro* bulbous protrusions; *V* vacuole; *single arrows* myelin-like figures. (After Batanouny et al. 1990)

12.3.1.2
Ecological Factors Affecting Salt Secretion

The rate of salt excretion is affected by numerous factors. These include: the salt concentration of the growth medium, the ions that are present, light, temperature, oxygen pressure, the presence of metabolic inhibitors and ratios between essential and toxic ions all of which can become limiting. Changes in the Na^+/K^+ ratio may determine the activity of several key metabolic processes. Selective secretion of Na^+ and Cl^- against nutritive ions prevents the domination of Na^+ and Cl^- over these ions and has been demonstrated in most species of salt-secreting plants (Waisel 1961; Atkinson et al. 1967; Rains and Epstein 1967).

12.3.2
Salt Bladders and Hairs

Salt secretion by bladders was discovered by Berger-Landefeldt (1959). The bladder trichomes were originally considered to be water storage organs (Wood 1923). Adriani (1958) called *Atriplex* a "non-desalting" halophyte. Since 1959, numerous studies have been conducted on this subject. Bladders occur in some genera of the Chenopodiaceae, essentially in all *Atriplex*

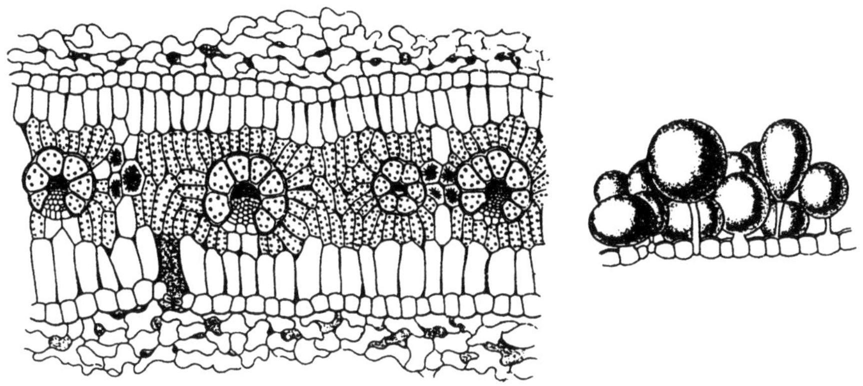

Fig. 68. A transverse section in the leaf of *Atriplex halimus* showing the dead bladders on the leaf surfaces. *Right* The salt bladders of *A. leucoclada* (Arabic: *Rughl*). (Volkens 1887)

species (Fig. 68), but also in some species of *Salsola,* in most species of *Chenopodium,* and in all species of *Obione* and *Halimione.* Figure 68 shows a transverse section in the leaf of *Atriplex halimus* with the dead bladders covering both surfaces of the leaf (Volkens 1887). It also shows the salt bladders on the leaf of *Atriplex leucoclada.* Volkens (1887) mentioned that the bladders of *A. leucoclada* are filled with water. There is a great similarity in the appearance of all swollen epidermal hairs or vesicles. The vesicle is usually stalked (1-4 stalk cells) and bears a swollen bladder cell, 80–200 μm in diameter. The structure of the stalk cell is similar to that of various cells of the salt glands. It is composed of thick cytoplasm rich in mitochondria, a dense endoplasmic reticulum, and numerous small vesicles. However, unlike salt glands, salt hairs contain chloroplasts. Stalk cells are connected to adjacent bladder or mesophyll cells by numerous plasmodesmata. The presence of salt hairs is considered to be an adaptation to salinity. The elimination of salts from the leaves of *Atriplex* into the vacuoles of epidermal bladder cells very effectively reduces the salt load in the metabolically active photosynthesising tissue of plants growing in highly saline media. In desert annuals (e.g. *Atriplex spongiosa*) that thrive for only a few weeks after seed germination, this elimination of salt into the bladder vacuoles may be sufficient for the entire life span of a leaf. In perennials such as *Atriplex nummularia,* collapsing and drying salt hairs eventually form a highly reflective silvery salt crust on the leaves that is periodically washed away by rain. New salt hairs are formed continuously. It seems that perennial species have better adaptive mechanisms than annual ones, at least with respect to the accumulation of chlorides. In *Halimione portulacoides,* Baumeister and Kloos (1974) found that the secreted sap (bladder cell sap) always contains a higher concentration of ions at a high salinity level than the sap pressed out of mesophyll cells. A powerful salt pump must operate in the stalk cell in order to maintain and produce such a high salt gradient between the bladder and the adjoining leaf cells. Mozafar and Goodin (1970) found a 60-fold difference in the accumulation of salt in between the bladders and the mesophyll cells. Bladders of halophytic *Atriplex* (Fig. 68) and *Halimione* species help them to cope with high salt concentrations. However, the importance varies from one species to the other. The induction of succulence under saline conditions in some species possessing bladder hairs demonstrates that additional factors must be considered when discussing the salt tolerance of such species. Osmond et al. (1980) stated that: "ion secretion to the bladder cell of epidermal trichomes functions to control concentration in the leaf lamina. When this control fails or if the species is poorly endowed with trichomes, a measure of regulation of salt concentration can be achieved by increased succulence. The latter response presumably involves a complex interaction between ion uptake, cell water relations, and cell wall metabolism."

12.4 Shedding of Salt-Saturated Tissues and Organs

This is considered to be one of the most important mechanisms for removing large quantities of salt from the bodies of halophytes. In *Juncus* species such as *J. rigidus* (syn. *J. maritimus*) and *J. rigidus* var. *arabicus* (syn. *J. arabicus*), the leaves are shed after being loaded with undesirable ions. The death and shedding of old leaves help the plant to eliminate considerable amounts of salts that were absorbed during its life span (Fig. 69). It is significant that the death of the leaves starts from the tip and proceeds downwards because the amount of salt contained in the leaves is usually higher at the tips than at the bases. In the case of articulate leafless chenopods, only some of the branches are shed. In addition to this partial death for the sake of survival, the fleshy cortex is shed from the stems of old branches. This occurs through the formation of phellogen deep in the cortex; the cork layers thus produced push against the fleshy cortex, which is finally discarded. The tissues of this succulent cortex with its fleshy cells accumulate considerable amounts of salt during the life of the plant. The older the plant, the more concentrated is the salt in its cortical tissues. Shedding of the cortex means discarding undesired salts from the plant. It also results in the removal of photosynthesising tissue along with the stomata on the epidermis. Shedding of the cortex is therefore also a mechanism by which the plant reduces its transpiring surface This is a useful mechanism in halophytes growing under arid conditions. Shedding of the cortex has been observed in *Arthrocnemum macrostachyum*. (Fig. 70), *Halocnemum* spp., and *Salicornia* spp. Analysis of

Fig. 69. *Juncus rigidus*in the saline areas of Ras El-Khaima (United Arab Emirates) showing the dead peripheral parts of the plant

the cortex of *Zygophyllum album* at different stages showed that the young cortex has a total soluble salt content of 16.0 %; the mature cortex contains 19.2 %, while the shed cortex contains 31.1 % (Ismail 1998). The chloride contents in the three successive stages were 11.0, 11.6, and 21.6 %, respectively.

In halophytes with succulent leaves, e.g. *Salsola* and *Suaeda* spp., the leaves are shed after being loaded with salt. Ismail (1998) found that the salt content of the shed leaflets of *Z. album* reaches 17.0 %, while young leaflets contain only 14.2 %. The water content of these leaflets was found to be 75.2 % in green leaves and 58.5 % in shed leaves (Ismail 1998). According to Chapman (1968), succulents such as *Halocnemum*, or *Salicornia* discard portions of their fleshy cortex together with their leaves, thus releasing large quantities of salts from the plants and enabling them to survive. The old

Fig. 70. Decortication of the branches of *Arthrocnemum macrostachyum*

Fig. 71. A branch of *Suaeda vermiculata* with black dead leaves

leaves of *Suaeda vermiculata* become black after being loaded with a large amount of salt and are shed, releasing considerable amounts of salt (Fig. 71). The generic name *Suaeda* came from the Arabic name *soweid* denoting the black colour of the old leaves (Batanouny 1981a, 1985, 1986a). This, in turn, was derived from the Arabic word *aswad* meaning black. In *Tamarix nilotica*, some of the twigs are shed and dispose excess salts. It was found by Ismail (1998) that the mean annual total soluble salts content of the shed twigs reached 21.5 %, while it was only 5.9 % in the twigs.

12.5 Osmotic Adjustment

Plants growing under saline conditions must maintain a high concentration of osmotically active substances in order to successfully counteract the water-retaining capacity of the surrounding medium (Waisel 1972). The importance of osmotic adaptation in terms of ion uptake is that in most halophytes the free energy of water is lowered by a massive increase in the ion content of the plant, which can reach very high values. Flowers (1975) gave examples showing that in some species of the genera *Suaeda, Salicornia* and *Atriplex*, 20–40 % of the dry weight may be due to inorganic salts. The osmotic potentials of halophytes are significantly lower than those of glycophytes, and greater negative osmotic potentials are obtained in halophytes than in glycophytes. The increased negativity of the osmotic potential of cell sap associated with osmotic adjustment to salinity is partly accomplished by the accumulation of ions and under certain conditions by large amounts of organic acids (Bernstein 1961). However, the major components of the salt content are Na^+ and Cl^- ions. In an extensive survey of the literature, Arnold (1955) deduced that chlorides accounted for between 35 and 95 % (average 65 %) of the osmotic potential of halophytes, between 15 and 65 % (average 35 %) of marsh glycophytes, and between 0 and 65 % (average 17 %) of other glycophytes (see Flowers 1975). The large contribution of inorganic ions to the osmotic potential of halophytes is another difference between halophytes and glycophytes (Onal 1965, 1966). In some species of halophytes, e. g. *Atriplex* spp., organic ions may appreciably increase the cellular osmotic potential (Strogonov 1964). Osmotic adaptation of plants such as *Artemisia salina* and *Anabasis salsa* results mostly from the accumulation of oxalates (Strogonov 1964). These play an important role as counter anions, especially in the chenopods found in semi-arid and arid zones. Oxalate levels increase with increasing Na^+ concentration in the medium (Williams 1960). Na^+/K^+ ratios are generally higher in halophytes than in glycophytes, although there is a relative enrichment of K^+ in the halophyte plant compared to in the medium (Flowers 1975). The enrichment of K^+ in the tissues relative to the medium is indicative of the ability of halophytes to absorb essential elements in the presence of vast excesses of other elements It has been suggested that the distribution of ions within the cells is discontinuous, i.e. that the "soluble phase" of the cytoplasm is relatively low in ions, while the vacuoles contain the majority of the absorbed ions (Flowers 1975). Waisel (1989) claimed that compartmentalisation of Na^+ and Cl^- inside cell vacuoles is the major mechanism that enables normal metabolic activity in high-salt cells to take place. This claim was based on the observation that many extracted plant enzymes lost their activity upon subjection to NaCl in vitro. However, this hypothesis could not be generalised, since the cytoplasm of various cells of salt-affected plants contains large quantities of Na^+ and Cl^-. Furthermore, the activity of various enzymes previously thought to be salt-sensitive can be altered so as to become salt-tolerant or even salt-stimulated (Waisel 1989). Osmotic adaptation differs not only between species but also between organs and even between tissues and cells. Osmotic adaptation is not a static phenomenon; periodic changes occur in the osmotic potential of plants.

12.6 Succulence

Exposure of halophytes to salinity results in numerous structural changes. As described by numerous investigators, they include: thicker leaves, larger cells (especially those of the spongy parenchyma), smaller intercellular spaces, greater elasticity of the cell walls, fully developed water-storing tissues, a smaller relative surface area (surface/volume ratio), low chlorophyll content and smaller and fewer stomata per surface area (Stocker 1933).

It is thought that succulence is one of the major factors in salt tolerance. Increased succulence has a diluting effect on the toxic ion content of cells, thus enabling plants to withstand the effects of larger quantities of salts. Waisel (1972) stated that this system has disadvantages due to the consequent lowering of nutritional ionic content. Two types of succulent halophytes have been distinguished on the basis of their water relationships: succulent mesomorphs and succulent xeromorphs (Weissenbock 1969). In the first type – typical of hydrohalophytes – all leaf cells, including epidermal cells, are succulent and the number of stomata per unit area is low (Weissenbock 1969; Biebel and Kinzel 1965). Succulent and large mesophyll cells, an enlarged cortex, small epidermal cells and abundant stomata characterise xeromorphs, which usually grow in habitats where the water stress is high. These plants also have a highly developed and highly lignified vascular system. The cell sap of succulent xeromorphs usually contains more organic acids than does the sap of succulent mesomorphs, which contains more inorganic ions.

The well-known work of Van Eijk (1939) showed that sodium chloride is the most efficient salt in promoting succulence and that chloride rather than sodium plays the major role in its induction. Jennings (1968) stated that halophytes differ from glycophytes in their response to sodium as a factor inducing succulence, but only in degree, not in kind. It is to be noted that the effect of sodium parallels that of increased light and aridity. Jennings (1968) proposed that these three factors exert the same effect on plant cells, namely to bring about increased synthesis of ATP. Succulence is caused by the shift in ATP metabolism induced by sodium transport. The ability of sodium to bring about increased succulence is two-fold. First, increased succulence has a diluting effect on the ion content of the cells, which might otherwise rise to toxic levels. Second, sodium acts as a specific stimulant of growth, which tends to reduce the turgor pressure component of the water potential of the cell. This growth response increases the water potential of the leaf cells without the plant having necessarily to absorb more ions. The action of sodium can thus be envisaged as eliciting a homeostatic response within the plant to this otherwise toxic ion such that its toxic action tends to be nullified (Jennings 1968).

A decrease in succulence, together with an increase in xeromorphic characteristics, has been observed in plants exposed to sodium sulphate. Chapman (1974) stated that there is a definite relationship between the concentration of Cl^-/SO^{4-}, and the degree of succulence in typical halophytic succulents, the Cl^-/SO^{4-}, ratio being strongly in favour of the Cl^- ions. Strogonov (1964) confirmed that the relative proportion of Cl^- and SO^{4-} ions is important. The former promotes the development of halosucculent characteristics, the latter of haloxeric features.

12.7 Proline and Other Organic Metabolites

Accumulation of compatible solutes under salt stress was at one time thought to be a basic process of adaptation, especially among plants that do not accumulate inorganic ions for osmotic adjustment. Though amino acids accumulate in all plants to some extent, proline accumulates consistently in certain species and under a wide range of environmental conditions. The metabolic effects of wilting that lead to proline accumulation have not been observed in the case of any other amino acid (Stewart 1981). Stewart and Lee (1974) highlighted the importance of proline to halophytes. These authors found that proline accounts for more than 30% of the amino acid pool in most halophytes (the only exception among 11 species tested is *Plantago maririma).* Batanouny et al. (1985) found that the diurnal value of proline in the twigs of the halophytic grass *Sporobolus pungens* (syn. *S. virginicus*) ranged from 20 μmol/g at 06.00 h to 45.2 μmol/g at 17.00 h (Fig. 72). Water conditions in photosynthesising (transpiring) organs are reflected in proline content and its accumulation. Batanouny et al. (1985) stated that, although proline accumulates at various ecological amplitudes as a response to water stress, in the tissues of different plants, the amount of proline could not be taken as a diagnostic feature of plants from different ecological groups. Negative relationships have been reported between salt tolerance and proline accumulation (Mendelssohn 1987).

12.8 Photosynthetic Pathways

Both C_3 and C_4 photosynthetic pathways occur in halophytes. While many succulent chenopods are C_4 plants, e.g. *Suaeda* spp., and *Seidlitzea rosmarinus*, there are other succulent C_3 chenopods, e.g. *Arthrocnemum* sp., *Halocnemum* sp. and *Halopeplis* sp., that do not use the C4 pathway.

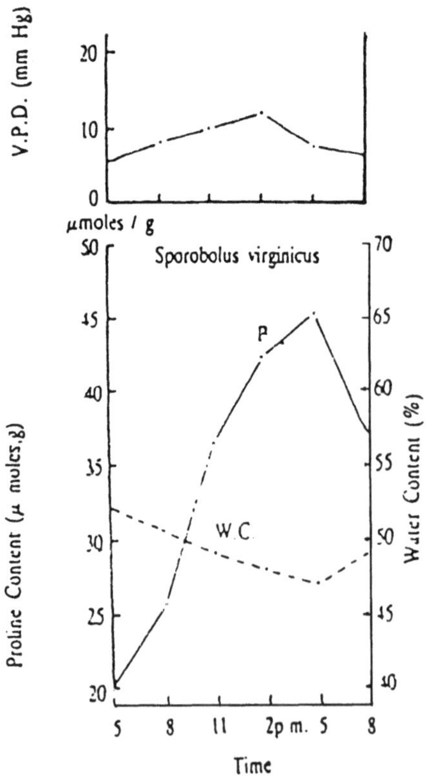

Fig. 72. Diurnal changes of proline (*P*) and water content (*W.C.*) of *Sporobolus pungens*. The change in vapour pressure deficit (*V.P.D.*)during the experiment is indicated. (After Batanouny et al. 1985)

Table 8 presents a list of C_3 and C_4 halophytes that grow at different salinities in the Middle East (Frey and Kürschner 1983; Frey et al. 1985). Halophytic grasses belonging to the subfamilies Panicoideae and the Chloroideae are salt-resistant C_4 plants (Liphschitz and Waisel 1982). The salt tolerance of some of these species is achieved by salt excretion. The presence of both photosynthetic pathways in halophytic plants may be evidence that the type of photosynthetic pathway is not correlated with tolerance of or resistance to salinity. However, Laetsch (1974) stated that the primary adaptation of C_4 plants was probably to saline environments and that the structural and functional adaptations are a prerequisite for efficient assimilation and growth. The photosynthetic spectrum studied by Frey et al. (1985) showed a characteristic zonation of C_3 and C_4 species along a transect in littoral salt marshes in different places in southwest Asia where moisture and salinity are the determining factors. The formations in which the C_4 pathway dominates are subject to stress with regard to both soil moisture and salinity. These are mainly xerohalophytes. By contrast, the halophytes in which the C_3 pathway is prevalent only have to withstand salt stress. These are the hygrohalophytes and mesohalophytes represented

Table 8. A list of C_3 and C_4 halophytes growing in different saline habitats in the Middle East

C_4 halophytes
Aleuropus lagopoides (L.) Thin ex Thw.
A. littoralis (Gouan) Pari.
Atriplex farinosa Forssk.
Halopyrum macronatum (L.) Stapf.
Seidlitzia rosmarinus Bunge ex Boiss.
Sporobolus arabicus Boiss.
S. spicatus (Vahl) Kunth.
Suaeda fruticosa Forssk. ex Gmel.
S. monoica Forssk.
S. pruinosa Lange
S. vermiculata Forssk. ex Gmel.
C_3 halophytes
Arthrocnemum macrostachyum (Moris) Moris Deplonte
Avicennia marina (Forssk.) Vierh.
Cressa cretica L.
Frankenia pulverulenta L.
Halocnemum slrobiolaceum (Pall.) M.B.
Halopeplis perfoliata (Forssk.) Bge ex Schw.
Limoneum axillare (Forssk.) Ketze
L pruinosum (L.) Ketze
Nitraria retusa (Forssk.) Aschers.
Zygophyllum album L.

in the study of Frey et al. (1985) on the vegetation of the *Avicennia* and *Halopeplis* zone. Such plants achieve their salt tolerance by means of different adaptations. Both types of stress (salt stress and water stress) must apparently occur together if there is to be a selection advantage for the C_4 pathway (Frey and Kürschner 1983). Although it is recognised that some halophytes require Na^+ as a micronutrient, it has nevertheless been shown that this requirement is restricted to C_4 plants (Brownell and Crossland 1972). In most respects the effects of salinity on photosynthesis are mediated via the water relationships and associated processes of the leaves. It is therefore not surprising that induction to crassulacean acid metabolism (CAM) in succulent annuals such as *Mesembryanthemum crystallinum* in response to NaCl can also be brought about by any treatment that reduces water uptake, such as growth in nutrient solutions at low temperature or low osmotic partial pressure (Winter 1979). Although succulent annuals are most commonly found in saline habitats that are seasonally arid, the change in CO_2 fixation is a response to water stress, not to salinity per se (Osmond et al. 1982).

12.9 Germination and Vegetative Reproduction

The salt tolerance of a plant at germination does not always imply tolerance in other growth phases. However, Waisel (1958) noted a positive correlation between salt resistance during germination and the salinity of the respective habitat in the case of some halophytes. He found that plants growing naturally in sites with a high salt content show a higher resistance to salt at the time of germination than do plants from less salty environments. A similar correlation between the ability to germinate under saline conditions and the zonation of plants in a salt marsh was reported by Ungar (1965, 1967). It has been observed that some halophytic species are capable of germination in salinities beyond those tolerated by salt plants (Waisel 1972); however, not all halophytic species are able to germinate in their native habitats without earlier leaching of the soil. In this respect it is significant that control of germination is sometimes achieved by the salt content of the disseminule. The seed germination in halophytic species of the genus *Atriplex* is controlled by the chloride content of the bracts enveloping the seeds (Beadle 1952) and by other osmotic effects. A similar phenomenon has been reported in the xerohalophyte *Zygophyllum dumosum*, the germination of which is regulated by the salt content of the dispersal unit. Salinity not only has a remarkable effect on percentage germination, but it also slows down the process. Extending the germination process over a lengthy period may be very critical to a plant's life, especially under arid conditions. The water relations of their habitats affect the germination of halophyte seeds. Germination of seeds of different species in the same habitat does not occur at the same time; instead there is a temporal distribution throughout the season. In addition to the physiological adaptations, the morphological aspects of seeds and seed germination may also contribute to the success of the plants (Waisel 1972). However, the contribution of polymorphism in the seeds of halophytes to the adaptability of their parent plants is still obscure. Many species of *Atriplex* are known to bear polymorphic dispersal units. For instance, A. *hortensis* has four kinds of flowers, each giving rise to fruits of different colours and shapes. The seeds of *A. dimorphostegia* also bear dispersal units of different size that vary in their germination behaviour. Even in similar dispersal units, the seeds of *Atriplex* exhibit polymorphism and vary in germination (Beadle 1952). The germination of seeds collected from different ecotypes of the same taxa is variable. Waisel (1972) found that seeds collected from halophytic populations of *Limonium pruinosum, Alhagi greacorum, Prosopis farcta* (syn. *Lagonychium farctum*), and *Phragmites australis* (syn. *P. communis*) germinate better under saline conditions than do seeds from glycophytic ecotypes of the same species. He stated that generalisation about the matter is still speculative as different modes of adaptation can be found among various ecotypes within one species.

The germination of mangrove species deserves particular attention, since problems in the germination of their seeds under conditions prevailing in the mangrove habitat have to be overcome by special mechanisms and adaptations. Seeds of those species that fall into the mud might sink and could suffocate. In most mangrove species, seed germination occurs immediately after the seeds mature, with no intervening period of dormancy. Many mangrove species are viviparous, e.g. *Rhizphora mucronata, Bruguiera* and *Avicennia* spp. It is interesting that in *Rhizophora* (Arabic: *quondol*), the seedling develops a long hypocotyl (20–40 cm) while the seedling is still attached to the mother plant. This ensures anchorage of the seedling when it falls and that the germinating seedling has a supply of light and oxygen; it also prevents it from being washed away. Seedlings of most species of mangrove withstand short periods of inundation.

Vegetative reproduction may in many cases be considered as well-adapted in halophytes. This is true for various species that are restricted to narrow ecological amplitudes, either by the production of disseminules or by their germination. Waisel (1972) concluded that in species such as *Aleuropus littoralis, Prosopis farcta* and *Tamarix aphylla,* vegetative reproduction seems to have an advantage over reproduction by means of seeds. In the case of *Aleuropus,* growth of the developing roots has the advantage of rapid penetration through the upper salty soil layer. Waisel and Pollak (1969) found that roots developing at different places along the runners are not dependent on a direct supply of water from the soil. As they are well supplied with water by the parent plant, their roots can penetrate layers of extreme salinity. Vegetative reproduction is much less affected than germination by salinity and temperature. It is the major means of reproduction for *Limonium vulgare* and *L. humile.* Cuttings of *Tamarix aphylla* are hardy and can withstand extreme environmental conditions. The ability of plants such as *Arthrocnemum, Nitraria, Limoniastrum, Salicornia* and *Limonium* spp. to form adventitious roots on their twigs is an important adaptation. It is helpful in vegetative production and may be the only means of reproduction in large parts of the area of distribution of some species.

12.10 Formation of Phytogenic Mounds

Numerous species of desert and marsh plants have the ability to build mounds around their shoots. The formation of phytogenic hillocks depends on numerous factors, including the habit of the plant, its life form, rooting habit, and environmental factors (Batanouny and Batanouny 1968). Successive stages of hillock formation may often be observed in the same locality. Plants in the seedling stage grow at ground level without any sand or soil accumulation. The growth of the plant helps the accumulation of deposits (wind- or water-borne materials), and a hillock is formed, rising above the ground level. Such mounds are augmentative to a certain limit, depending on a number of ecological factors. The formation of mounds has many consequences. The accumulation of deposits around the shoot system of the plant stimulates the development of adventitious roots within the confines of the mound. The newly formed adventitious roots are of different types and shapes, with varied ramifications and extensions. They may be morphologically similar to the laterals produced by the primary root, as in *Arthrocnemum macrostachyum* (syn. *A. glaucum*) and *Suaeda pruinosa* (Batanouny and Batanouny 1969). In other plants, such as *Anabasis articulata* (a xerophyte with halophytic affinity) and *Z. album,* the adventitious roots are hairy, very fine, and short, and their extension is confined to the interior of the mound body. In a third group, distinguished by Batanouny and Batanouny (1969), both the long thread-like and the fine tufted roots are produced from the old buried stems, as seen in *L. monopetalum* (Figs. 22, 73). The newly produced adventitious roots may have larger absorbing surfaces than the primary root system. They are responsible for a considerable part of the total water intake of the plant. The fine, short roots, produced in tufts by *L. monopetalum,* are considered by Batanouny and Abu Sitta (1977) to be ephemeral roots. They are produced during the wet season and vanish and lose their activity with the advent of summer. This type of root does not extend below the body of the mound. The conditions created by the

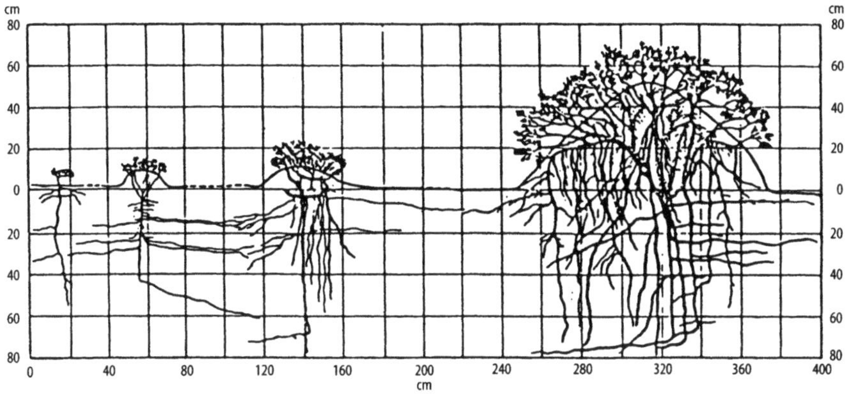

Fig. 73. The root system of *Limoniastrum monopetalum* during successive stages of mound formation by the plant. (After Batanouny and Abu Sitta 1977)

formation of the mound may be of help in the condensation of internal dew, supplying the roots with fresh water, and thereby diluting the soil solution from dawn to a few hours after sunrise. The elevation of a mound above the level of the ground surface may cause rapid cooling at night, particularly at its surface. This leads to condensation of both external and internal dew from the atmospheric moisture. The latter occurs when the temperature gradient is favourable for an upward movement of water vapour from the deep warm layers below. Formation of mounds modifies the soil properties. The soil that forms the mound is usually more penetrable and looser than that between mounds. In salt marshes, the crust of salts on the surfaces of the mounds is thinner than the one covering the ground between them. The salt contents within a mound is different than outside it, but one cannot conclude which is higher and in what soil layer, except after a thorough study of the mounds. The changes in the physical properties of the soil, and consequently in the soil-water relationships, initiate the formation of adventitious roots. Their development from the buried vegetative parts at the subsurface level may be indicative of the importance of the water from external dew that condenses on the surface and in the body of the mound.

High salinity has an adverse effect on the rooting of halophytes. Batanouny and Abu Sitta (1977) found that *L. monopetalum* grows in a wide variety of habitats in which the salt contents range from 13.3 to 25.5 % in highly saline habitats and from 1.1 to 3.5 % in less saline regions. High salinity hinders the production of adventitious roots. Furthermore, the primary root system is confined to the soil layers above the water table. Together, these conditions of high salinity and a shallow water table result in stunted, prostrate *L. monopetalum* individuals (Fig. 74).

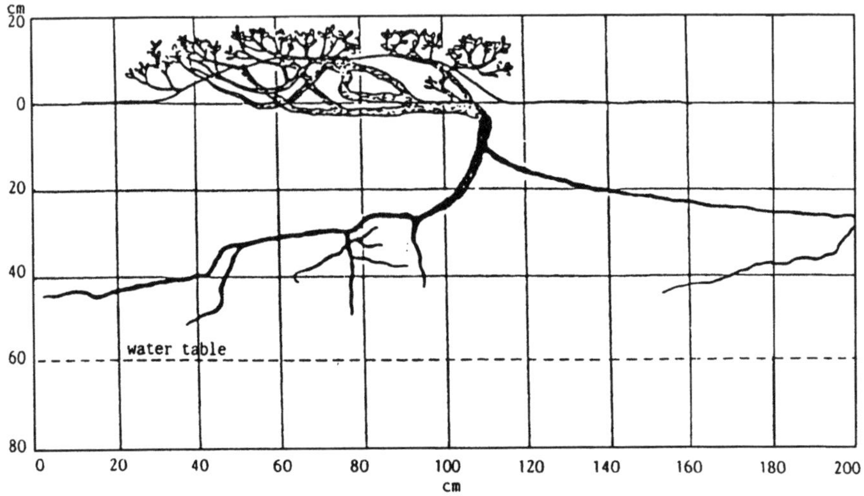

Fig. 74. The root system of *Limoniastrum monopetalum* growing under highly saline conditions and with a shallow water table. (After Batanouny and Abu Sitta 1977)

The above-mentioned adaptations and mechanisms help the different halophytes to grow under salinity stress. The equilibrium between these various mechanisms is a delicate one. Each species has its own array of adaptive mechanisms. The ability of a plant to survive, grow and reproduce, even survive without net growth, means that such a plant is endowed with particular adaptations enabling it to resist and/or tolerate salinity.

References

Aamodt OS, Johnston WH (1936) Studies on drought resistance in spring wheat. Can J Bot 14:122–152

Abdel Rahman AA, Batanouny KH (1959a) Seasonal variation in the desert vegetation along Cairo-Suez road. Bull Inst Désért d'Égypte 9:1–10

Abdel Rahman AA, Batanouny KH (1959b) The phenology of desert plants in relation to environment. Bull Inst Désért d'Égypte 9:11–19

Abdel Rahman AA, Batanouny KH (1959 c) Germination of desert plants under different environmental conditions. Bull Inst Désért d'Égypte 9:21–40

Abdel Rahman AA, Batanouny KH (1959d) Root development and establishment under desert conditions. Bull Inst Désért d'Égypte 9:41–50

Abdel Rahman AA, Batanouny KH (1964a) Transpiration of desert plants under different conditions. J Bot (UAR) 7:37–5

Abdel Rahman AA, Batanouny KH (1964b) Osmotic pressure of desert plants under different environmental conditions. J Bot (UAR) 7:95–107

Abdel Rahman AA, Batanouny KH (1965a) The water output of the desert vegetation in the different microhabitats in wadi Hoff. J Ecol 53:139–145

Abdel Rahman AA, Batanouny KH (1965b) Transpiration of desert plants under different environmental conditions. J Ecol 53:267–272

Abdel Rahman AA, Batanouny KH (1965 c) Vegetation and root distribution of plants in the different microhabitats of wadi Hof. Bull Inst Désért d'Égypte 15:55–66

Abdel Rahman AA, Batanouny KH (1966) Microclimatic conditions in wadi Hoff. Bull Soc Geogr d'Égypte 39:137–153+l pl

Abdel Rahman AA, El Hadidy MN (1958) Observations on water output of the desert vegetation along Suez Road. Egypt J Bot 1:19–38

Abdel Rahman AA, El-Shourbagy MN, Shalaby AF, El-Monayaeri MO (1972) Salinity effects on growth and water relations of some desert range plants. Flora 161:495–508

Addoms RM (1946) Entrance of water into suberized roots of trees. Plant Physiol 21:109–111

Adriani MJ (1958) Halophyten. In: Ruhland W (ed) Handbuch der Pflanzenphysiologie, vol 4. Springer, Berlin, Heidelberg, New York, pp 709–736

Aerts R, Boot RGA, van der Aart PJM (1991) The relation between above- and belowground biomass allocation patterns and competitive ability. Oecologia (Berlin) 87:551–559

Akhatar P, Shaukat SS (1976) Role of dew in the survival and phenology of *Gossypium hirsutum* L. cv. Qalandri. Pak J Bot 8:151–155

Akhatar P, Shaukat SS (1979) Drought endurance and dew utilization capacity of *Sorghum bicolor* (L.) Moench and *Ipomoea pes-caprae* (L.) Sweet. Pak J Bot 11:85–91

Alpert P (1979) Desiccation of desert mosses following a summer rainstorm. Bryologist 82: 65–71

Amundson R (1990) Doctor Dennet and Doctor Pangloss: perfection and selection in biology and psychology. Continuing commentary. Behav Brain Sci 13:577–584
Amundson R (1994) The comcepts of constraint: adaptation and the challenge from developmental biology. Philos Sci 61:556–578
Amundson R (1996) Historical development of the concept of adaptation. In: Rose MR, Lauder GV (eds) Adaptation. Academic Press, San Diego, pp 11–53
Amundson R, Lauder GV (1994) Function without purpose. The use of casual role function in evolutionary biology. Biol Philos 9:443–469
Arisz WH, Camphuis IJ, Heikens H, Van Tooren AJ (1955) The secretion of the salt glands of *Limonium latifolium* Ktze. Acta Bot Neerl 5:218–246
Arnold A (1955) Die Bedeutung der Chlorionen für die Pflanze insbesondere phsiologische wirksam. Bot Stud, Teil 2. Fischer, Jena
Atkinson MR, Findlay GP, Hope AB, Pitmaii MG, Saddler HDM, West KR (1967) Salt regulation in the mangroves Rhizo*phora mucronata* Lam. and *Aegialitis annulata* R. Br Aust J Biol Sci 20:589–599
Babu R, Went FW (1978) The effect of dew on plant water balance in desert shrubs. An Arid Zone 17:1–11
Barbour MG, Pavlik BM, Drysdale FR, Lindstorm S (1993) California's changing landscapes. California Native Plant Society, Sacramento
Barnett NM, Naylor AW (1966) Amino acid and protein metabolism in Bermudagrass during water stress. Plant Physiol 41:1222–1230
Baskin CC, Baskin JM (1974) Responses of *Astragalus tennesseensis* to drought. Changes in free amino acids and amides during water stress and possible ecological significance. Oecologia 17:11–16
Batanouny KH (1960) Serasonal variations in the desert vegetation along Cairo-Suez desert road. MSc Thesis, Faculty of Science, Cairo University
Batanouny KH (1963) Water economy of desert plants in wadi Hoff. PhD Thesis, Faculty of Science, Cairo University
Batanouny KH (1973a) Soil properties as affected by topography in desert wadis. Acta Bot Acad Sci Hung 19:13–21
Batanouny KH (1973b) Kalkdrüsen von *Limoniastrum monopetalum.* Naturwissensch Rundsch 26:213–214
Batanouny KH (1974) Eco-physiological studies on desert plants. IX-Types of transpiration curves of *Zilla spinosa* Prantl under natural conditions. Flora 163:1–6
Batanouny KH (1975) Water resources and plant life in the Egyptian desert. In: Vansteekiste GC (ed) Proc IFIP Working Conference. Computer Simulation of Water Resources Systems. North-Holland, Amsterdam
Batanouny KH (1979a) Vegetation along Jeddah-Mecca road: pattern and process as affected by human impact. J Arid Environ 2:21–30
Batanouny KH (1979b) The desert vegetation in Egypt. Cairo Univ Afr Stud Rev Sp Publ 1:9–37
Batanouny KH (1980) Water economy of desert plants. In: Halasi-Kun GJ (ed) Pollution and water resources. Columbia Univ Seminar Series. Pergamon Press, Oxford, New York, pp 167–177
Batanouny KH (1981a) Ecology and flora of Qatar. Centre for Applied and Scientific Research, University of Qatar Doha, 2545 pp+124 coloured plates
Batanouny KH (1981b) Eco-physiological studies on desert plants. X-Contribution to the autecology of the desert chasmophyte *Stachysa aegyptiaca* Pers. Oecologia (Berl) 50:422–426
Batanouny KH (1983) Human impact on desert vegetation. In: Holzner W, Werger MJ, Ikusima I (eds) Man's impact on vegetation. Junk, The Hague, pp 139–149
Batanouny KH (1984) Rangelands of the Arabian Peninsula, with a special reference to the history of range management (The Hema, an old Arabian reserve system). Working paper 2nd Int Rangeland Congr. Adelaide, Australia, 13–18 May 1984

Batanouny KH (1985) Latin botanical names of Arabic origin. Bull Fac Human Soc Sci Univ Qatar 9:395–431 (in Arabic)
Batanouny KH (1986a) Plant life and environment in the State of Qatar. Univ Qatar Doha Qatar (in Arabic)
Batanouny KH (1986b) Plants in the Hadith of the Prophet. The directorate for the revival of islamic heritage, Qatar, 217pp+40 coloured plates (in Arabic)
Batanouny KH (1990) Rangeland ecology in the Arab Gulf countries. In: Halwagy R, Taha F, Omar S (eds) Advances in range management in arid lands. Kegan Paul International, London, in association with the Kuwait Foundation for the Advancement of Sciences, pp 33–55
Batanouny KH (1992a) Lexicon of plants and botanical terms. In: El Ghoneim A (ed) Dictionary of the Holy Qura'n, 1st edn. Kuwait Foundation for the Advancement of Sciences, Kuwait, 158 pp (in Arabic, translated into English and French)
Batanouny KH (1992b) Adaptation of plants to saline conditions in arid regions. In: Lieth H, Masoom Al (eds) Towards the rational use of high salinity tolerant plants, vol 1. Kluwer-Academic, Amsterdam, pp 387–401
Batanouny KH (1994) Halophytes and halophytic plant communities in the Arab region. In: Squires VR, Ayoub AT (eds) Halophytes as a resource for livestock and for rehabilitation of degraded lands, vol 1. Kluwer-Academic, Amsterdam, pp 139–163
Batanouny KH (1999) Wild medicinal plants of Egypt: an inventory for conservation and sustainable use. Academy of Scientific Research and Technology, National Committee of IUCN (International Union for Conservation), Cairo
Batanouny KH, Abdel Wahab MA (1973) Eco-physiological studies on desert plants. VIII. Root penetration of *Leptadenia pyrotechnica* (Forssk.) Decne in relation to its water balance. Oecologia (Berl) 11:151–161
Batanouny KH, Abu Sitta YM (1977) Eco-physiological studies on halophytes in and and semi-arid zones. 1. Autecology of the salt secreting halophyte *Limoniastrum monopetalum* (L.) Boiss. Acta Bot Acad Sci Hung 23:13–31
Batanouny KH, Baeshin NA (1983) Plant communities alomng Medina-Badr road across the Hejaz mountains, Saudi Arabia. Vegetatio 53:33–43
Batanouny KH, Batanouny MH (1968) Formation of phytogenic hillocks. I. Plants forming phytogenic hillocks. Acta Bot Acad Sci Hung 14:234–252
Batanouny KH, Batanouny MH (1969) Formation of phytogenic hillocks. II. Rooting habit of plants forming phytogenic hillocks. Acta Bot Acad Sci Hung 15:1–18
Batanouny KH, Ebeid M (1981) Diurnal changes in proline content of desert plants. Oecologia (Berl) 51:250–252
Batanouny KH, Hilli MR (1973) Phytosociological study of Ghurfa desert, Iraq. Phytocoenologia 1:223–249
Batanouny KH, Khalifa T (1970) Contribution to the autecology of *Urginea maritima* in Egypt. Phyton 14:41–53
Batanouny KH, Zaki MAF (1969) Root development of two common species in different habitats in the Mediterranean subregion in Egypt. Acta Bot Acad Sci Hung 15:217–226
Batanouny KH, Hassan AA, Zayed KM (1985) Proline accumulation in plants of different ecological groups as a response to water deficit. Qatar Univ Sci Bull 5:131–143
Batanouny KH, Stichler W, Ziegler H (1988) Photosynthetic pathways, distribution, and ecological characteristics of grass species in Egypt. Oecologia (Berl) 75:539–548
Batanouny KH, Stichler W, Ziegler H (1991) Photosynthetic pathways and ecological distribution of *Euphorbia* species in Egypt. Oecologia (Berl) 87:565–589
Batanouny KH, Hassan AA, Fahmy GM (1992) Eco-physiological studies on halophytes in arid zones. II. The salt secreting *Limonium delicatulum* (Gir) Ktze. Flora 186:105–116
Bates LS (1973) Rapid determination of proline for water stress studies. Plant Soil 39:205–207

Baumeister W, Kloos G (1974) Über die Salzsekretion bei *Halimione portulacoides*. Flora 163:24–56
Beadle NC (1952) Studies in halophytes. 1. The germination of the seed and establishment of the seedlings of five species of *Atriplex in* Australia. Ecology 33(1):49–62
Beaumont P, Blake GH, Wagstaff M (1976) The Middle East geographical study. Wiley, Chichester
Bender MM (1968) Mass spetrometric studies of carbon-13 variations in corn and other grasses. Radiocarbon 10:468–472
Berger-Landefeldt U (1959) Beitraege zur Oekologie der Pflanzen nordafrikanischer Salzpfannen. Vegetatio 2:1–48
Berner T (1974) Studies on the ecophysiology of the hypolithic algae from the flint stones of the Meshash formation in the northern Negev, Israel. Thesis, Jerusalem
Bernstein L (1961) Osmotic adjustment of plants to saline media. 1. Steady state. Am J Bot 48:909–918
Bertsch A (1966) CO_2-Gaswechsel und Wasserhaushalt der aerophilen Grünalge *Apatococcus lobatus*. Planta 70:46–72
Bewley JD, Krochko JE (1982) Desiccation tolerance. In: Lange OL, Nobel PS, Osmond CB, Ziegler H (eds) Physiological plant ecology II. Water relations and assimilation. Springer, Berlin Heidelberg New York, pp 325–378
Bewley JD, Tucker EB, Gwózdz EA (1974) The effect of stress on the metabolism of *Tortula ruralis*. In: Bieleski RL, Ferguson AR, Cresswell MM (eds) Mechanisms of regulation of plant growth. R Soc NZ Bull 12:395–402
Bhardwaj N, Gopal Brij (1979) Study of root system of *Tephrosia apollinea* and its survival value under arid conditions. Jpn J Ecol 29:229–233
Biebel R, Kinzel H (1965) Blattbau und Salzhaushalt von *Laguncularia racemosa* (L.) Gaertn. und andere Mangrovenblumen auf Puerto Rico. Oesterr Bot Z 112:56–93
Blake G, Dewdney J, Mitchell J (1987) The Cambridge atlas of the Middle East and North Africa. Cambridge Univ Press, Cambridge
Blum OB (1973) Water relations. In: Ahmadjian V, Hale ME (eds) The lichens. Academic Press, New York, pp 381–400
Boaz M, Plitmann U, Heyn CC (1994) Reproductive effort in desert versus mediterranean crucifers: the allogamous *Erucaria rostrata* and *E. hispanica* and the autogamous *Erophila minima*. Oecologia 100:286–292
Boyce SG (1954) The salts pray community. Ecol Monogr 24:29–67
Breckle SW (1986) Studies on halophytes from Iran and Afghanistan. II. Ecology of halophytes along salt gradients. Proc R Soc Edinb 89B:203–215
Brock TD (1975) The effect of water potential on photosynthesis in whole lichens and in their liberated algal components. Planta 124:13–23
Brown JF (1997) Effects of experimental burial on survival, growth, and resource allocation of three species of dune plants. J Ecol 85:151–158
Brownell PF, Crossland CL (1972) The requirement for sodium as a micronutrient by species having the C_4 dicarboxylic acid photo synthetic pathway. Plant Physiol 49:794–797
Caldwell MM (1974) Physiology of desert halophytes. In: Reinold RJ, Queen WH (eds) Ecology of halophytes. Academic Press, New York, pp 355–377
Cannon WA (1906) The effects of high relative humidity on plants. Torreya 2:21–25
Cannon WA (1911) The root habit of desert plants. Carnegie Inst Washington Publ 131:1–96
Carr Dj, Gaff DF (1961) The role of cell-wall in the water relations of leaves. Proceedings of the Madrid Symposimon plant water relationships in arid and semi-arid conditions. Arid Zone Res 16:117–125
Chapman VJ (1968) Vegetation under saline conditions. In: Boyko H (ed) Saline irrigation for agriculture and forestry. Junk, The Hague, pp 201–216
Chapman VJ (1974) Salt marshes and salt deserts of the world, 2nd edn (reprint). Cramer, Lehre

Chen D, Kessler B, Monselise SP (1964) Studies on water regime and nitrogen metabolism of citrus seedlings grown under water stress. Plant Physiol 39:379–386
Cloudsley-Thompson JL (1962) Bioclimatic observations in the Red Sea hills and coastal plains, as a major habitat for the desert locust. Proc Ent Soc Lond A 37:27–34
Cloudsley-Thompson JL (1996) Biotic intercations in arid lands. Springer, Berlin Heidelberg New York
Cohen D (1971) Maximizing final yields when growth is limited by time or by limiting resources. J Theor Biol 33:299–307
Cohen E (1976) The optimal timing for reproduction. Am Nat 110:801–807
Cook CW (1943) A study of the roots of *Bromus inermis* in relation to drought resistance. Ecology 24:16–18
Crawford RMM (1989) Studies in ecology, vol II: studies in plant survival: ecological case histories of plant adaptation to adversity. Blackwell, Oxford
Crick JC, Grime JP (1987) Morphological plasticity and mineral nutrient capture in two herbaceous species of contrasted ecology. New Phytol 107:403–414
Cutler JM, Rains DW, Loomis RS (1977) The importance of cell size in the water relations of plants. Physiol Plant 40:255–260
Danin A (1991) Plant adaptations in desert dunes. J Arid Environ 21:193–212
Danin A (1996) Plants of desert dunes. In: Cloudsley-Thompson JL (ed) Biotic interactions in arid lands. Springer, Berlin Heidelberg New York
Danin A, Orshan G (1990) The distribution of Raunkiaer life forms in Israel in relation to the environment. J Veg Sci 1:41–48
De Bary A (1877) Vergleichende Anatomie der Vegetationsorgane. Engelman, Leipzig
De Fraine E (1916) The morphology and anatomy of the genus *Statice* as represented at Blakenely Point. *1. Statice binervosa* G.E. Smith and *Statice bellidifolia* D.C. (=*S. reticulata*). Ann Bot 30:239–282
Draz O (1979) Revival of the *hema* system of range reserves as a basis for he Syrian range management program. Proc 1st Int Rangeland Congr, pp 100–103
Draz O (1980) Range and fodder crop development, Syrian Arab republic. FAO AG: DP/ SYR/68/011, pp 79–88
Dregne HE, Nan-Ting Chou (1992) Global desertification dimensions and costs. In: Dregne H (ed) Degradation and restoration of arid lands. Int Center for Arid and Semi-Arid Lands Studies, Texas Tech Univ, pp 249–282
Ehleringer J, Björkman O, Mooney HA (1976) Leaf pubescence: effects on absorptance and photosynthesis in a desert shrub. Science 192:376–377
Eickhmeier WG (1979) Photosynthetic recovery in the resurrection plant *Selaginella lepidophylla* after wetting. Oecologia 39:93–100
Evenari M (1937/1938) The physiological anatomy of the transpiratory organs and the conducting systems of certain lands of the wilderness of Judea. J Linn Soc Lond 51: 389–407
Evenari M (1953) The water balance of plants in desert condition. In: Desert research proceedings. Research Council of Israel, Special Publ. No 2, pp 266–274
Evenari M, Richter R (1937) Physiological-ecological investigations in the wilderness of Judea. J Linn Soc (Bot) 51:333–381
Evenari M, Shanan L, Tadmor N (1971) The Negev, the challenge of a desert. Harvard Univ Press, Cambridge, 345 pp
Evenari M, Schulze ED, Kappen L, Buschbom U, Lange OL (1975) Adaptive mechanisms in desert plants. In: Vernberg EJ (ed) Physiological adaptation to the environment. Intext Educ Publ, New York, pp 111–129
Fahmy GM (1986) Eco-physiological studies on some halophytes in the Mediterranean zone, Egypt. PhD Thesis, Fac Sci, University of Cairo

Fahn A (1963) The fleshy cortex of the articulated Chenopodiaceae. Maheshwari Comm vol. J Indian Bot Soc 42 A:39–45
Fahn A (1989) Plant anatomy, 3rd edn. Pergamon Press, Oxford, 544 pp
Fahn A, Cutler D (1992) Xerophytes. Encyclopedia of plant anatomy, vol III. Gebrüder Bornträger, Berlin, 176 pp
Fahn A, Werker E, Bass P (1986) Wood anatomy and identification of trees and shrubs from Israel and adjacent regions. Israel Academy of Sciences and Humanities, Jerusalem, 221 pp
Ferrar JF (1973) Lichen physiology: progress and pitfalls. In: Ferry DW, Baddeley MS, Hawksworth DL (eds) Air pollution and lichens. Athlone, London, pp 238–282
Ferrar JF (1976) The lichen as an ecosystem: observation and experiment. In: Brown DH, Hawksworth DL, Baily RH (eds) Lichenology: progress and problems. Academic Press, London, pp 385–406
Fisher WB (1978) The Middle East. A physical, social and regional geography, 7th edn. Methuen, London
Fitter AH (1991) Characteristics and functions of root systems. In: Waisel Y, Eshel A, Kafkafi U (eds) Plant roots. Marcel Dekker, NewYork, pp 3–25
Fitting H (1911) Die Wasserversorgung und die osmotischen Druckverhältnisse der Wüstenpflanzen. Z Bot 3:209–275
Flowers TJ (1975) Halophytes. In: Baker DA, Hall JL (eds) Ion transport in plant cells and tissues. North-Holland, Amsterdam, pp 309–333
Flower N (1986) The role of competition in plant communities in arid and semiarid regions. Annu Rev Ecol Syst 17:89–110
Freitas Helna MO (1997) Drought. In: Prasad MNV (ed) Plant ecophysiology. Wiley, New York, pp 129–149
Frey W, Kürschner H (1983) Photosyntheseweg und Zonierung von Halophyten an Salzseen in der Türkei, in Jordanien und im Iran. Flora 173:293–310
Frey W, Kürschner H (1991a) Lebensstrategien von terrestrischen Bryophyten in der Judaeischen Wüste. Bot Acta 104:172–182
Frey W, Kürschner H (1991b) Morphologische und anatomische Anpassungen der Arten in terrestridschen Bryophytengesellschaften entlang eines oekologischen Gradienten in der Judaeischen Wüste. Bot Jahrb Syst 112:529–552
Frey W, Kürschner H (1991 c) Das Fossombronio-Gigaspermuetum mourtii in der Judaeischen Wüste. Crypt Bot 2/3:73–84
Frey W, Kürschner H (1992) Bryosoziologische Untersuchungen in Jordanien: terrestrische und epilithsche Gesellschaften. Nova Hedwigia 54(3/4):355–378
Frey W, Kürschner H, Stichler W (1985) Photosynthetic pathways and ecological distribution of halophytes from four littoral salt marshes.(Egypt/Sinai, Saudi Arabia, Oman and Iran). Flora 177:107–130
Frey W, Herrnstadt I, Kürschner H (1990) Verbreitung und Soziologie terrestrischer Bryophytengesellschaften in der Judaeischen Wüste. Phytocoenoligia 19(2):233–265
Gaff DF (1980) Protoplasmic tolerance of extreme water stress. In: Turner NC, Kramer PJ (eds) Adaptation of plants to water and high temperature stress. Wiley-Interscience, New York, pp 207–229
Gaff DF, Carr DJ (1961) The quantity of water in the cell wall and its significance. Aust J Biol Sci 14:299–311
Gaff DF, Hallam ND (1974) Resurrecting desiccated plants. In: Bieleski RL, Ferguson AR, Cresswell MM (eds) Mechanisms of regulation of plant growth. R Soc NZ Bull 12:389–393
Gadgil M, Gadgil S (1979) Adaptive significance of the relation between root and shoot growth. J Indian Inst Sci 61:25–40
Galun M (1970) The lichens of Israel. The Israel Academy of Sciencs and Humanities, Jerusalem, 116 pp

Galun M, Garty J (1972) Lichens of north and central Sinai. Isr J Bot 21:243–354
Galun M, Reichert I (1960) A study of the lichens of the Negev. Bull Res Counc Isr 9D:127–148
Gannutz TP (1969) Effects of environmental extremes on lichens. Bull Soc Bot Fr Mem 1968, Colloq Lichens 1967, pp 169–179
Gates DM, Alderfer R, Taylor E (1968) Leaf temperature of desert plants. Science 159:994–995
Gindel I (1964) Attraction of atmospheric moisture by woody xerophytes I. Arid climates. Commonw For Rev 45:297–321
Gould SJ, Vrba ES (1982) Exaptatin – a missing term in the science of form. Paleobiology 8:4–15
Grime JP (1979) Plant strategies and vegetation processes. Wiley, Chichester
Grime JP, Cric JC, Rincon JE (1986) The ecological significance of plasticity. In: Jennings DH, Trewavas AJ (eds) Plasticity in plants. Cambridge University Press, Cambridge, pp 5–29
Gupta RK (1978) The physiology of the desiccation resistance in bryophytes: effect of desiccation on water status and chlorophyll a and b in bryophytes. Ind J Exp Biol 16:354–356
Gutterman Y (1993) Seed germination in desert plants. In: Cloudsley-Thompson JL (1996) Biotic interactions in arid lands. Springer, Berlin Heidelberg New York, 253 pp
Haberlandt G (1884) Physiologische Pflanzenanatomie. Engelmann, Leipzig
Haines RW (1951) Potential annuals of the Egyptian desert. Bull Inst Déśrt d'Égypte 1:103–118
Hammouda MA (1954) Studies in the water relations and transpiration of Egyptian desert plants. PhD Thesis, Faculty of Science, Cairo University
Hanson AD, Nelson CE, Everson EH (1977) Evaluation of free proline accumulation as an index of drought resistance using two contrasting barley cultivars. Crop Sci 17:720–726
Harborne JB (1977) Introuction to ecological biochemistry. Academic Press, London
Harris GB (1976) Water content and productivity of lichens. In: Lange OL, Kappen L, Schulze E-D (eds) Water and plant life. Problems and modern approaches. Ecological studies, vol 19. Springer, Berlin Heidelberg New York, pp 452–468
Harris D, Davy AJ (1987) Seedling growth in *Elymus farctus* after episodes of burial with sand. Ann Bot 60:587–593
Hartmann H (1979) Surface structures of leaves: their ecological and taxonomical significance in members of the subfamily Ruschioideae Schw. (Mesembryanthmaceae Fenzl.). In: Cutler DF, Hartmann H (eds) Scanning electron microscope studies ofd the lewaf epidermis in some succulent. Tropische und subtropische Pflanzenwelt, vol 28. Mainz Akad Wiss Lit, pp 31–55
Hassib M (1951) Distribution of plant communities in Egypt. Bull Fac Sci Fouad I Univ Cairo 29:5
Hébant C (1977) The conducting tissues of bryophytes. Kramer, Vaduz
Heide-Jorgensen HS (1978) The xeromorphic leaves of *Hakea suaveolens* R.B.: structure of epidermal cells, cuticle development and ectodesmata. Bot Tidsskr 72:227–244
Henckel RA (1964) Physiology of plants under drought. Annu Rev Plant Physiol 15:363–386
Henckel RA, Pronina ND (1968) Factors underlying the dehydration resistance in poikilo-xerophytes. Sov Plant Physiol 15:68–74
Hoffman GJ, Rawlins ST, Gaber MJ, Cullen EM (1971) Water relations and growth of cotton as influenced by salinity and relative humidity. Agron J 63:822–826
Huber B (1956) Temperatur pflanzlicher Oberflächen. In: Ruhland W (ed) Handbuch der Pflanzenphysiologie, vol 3. Springer, Berlin, Heidelberg, New York, pp 285–292
Ismail SA (1998) Ecological study of the plant life in *Wadi El Rayan*, Fayoum. Egypt. PhD Thesis, Fac Sci, Cairo Univ, Egypt
Iljin WS (1935) Die Lebensfähigkeit der Pflanzenzellen in trockenen Zustand. Planta (Berl) 24:742–745
Iljin WS (1953) Causes of death of plants as a consequence of loss of water: conservation of life in desiccated tissues. Bull Torr Bt Club 80:166–177
Jarvis MS (1963) A comparison between the water balance of species with contrasting types of geographical distribution in the British Isles. Br Ecol Soc Symp. In: Rutter AJ, Whitehead FH (eds) The water relations of plants. Wiley, New York, pp 289–312

Jennings DH (1968) Halophytes, succulence and sodium in plants – a unified theory. New Phytol 67:899–911
Johnson HB (1975) Gas exchange strategies in desert plants. In: Gates DM, Schmerl M (eds) Prespectives of biophysical ecology. Ecological studies, vol 12. Springer, Berlin Heidelberg New York, pp 49–120
Kappen L (1973) Response to extreme environments. In: Ahmadjian V, Hale ME (eds) The lichens. Academic Press, London, pp 311–380
Kappen L, Lange OL, Schultze ED, Evenari M, Buschbom U (1975) Primary production of lower plants (lichens) in the desert and its physiological basis. In: Cooper JP (ed) Photosynthesis and productivity in different environments. Cambridge Univ Press, Cambridge, pp 133–143
Kappen L, Lange OL (1972) Die Kälteresistenz einige Makrolichenen. Flora 161:1–29
Kassas M, Batanouny KH (1984) Plant ecology in Sahara desert. In: Cloudsley-Thompson J (ed) Sahara desert. Pergamon Press, Oxford, pp 77–90
Kassas M, Imam M (1959) Habitat and plant communities in the Egyptian desert. IV. The gravel desert. J Ecol 47: 287–310
Kausch W (1955) Saugkraft und Wassernachleitung im Boden als physiologische Faktoren unter besonderer Berücksichtung des Tensiometers. Planta (Berl) 45:217–263
Kausch W (1965) Relation between root growth, transpiration and CO2-exchange in several cacti. Planta (Berl) 66:229–238
Kausch W (1968) Das Wurzelwerk der Pflanzen als Organ für die Wasseraufnahme. Umsch Wiss Tech 2:37–44
Kemble RA, MacPherson HT (1954) Liberation of amino acids in perennial ryegrass during wilting. Biochem J 58:46–49
Kerr JP, Beardsell MF (1975) Effect of dew on leaf water potentials and crop resistances in a paspalum pasture. Agron J 67:596–599
Killian CH (1941) Sols et plantes indicatrices dans les parties non irrigées des oasis de Figuig et de Beni-Ounif. Bull Soc Hist Nat Afr N 32
Killian CH, Faurel L (1933) Obervations sur la pressionosmotique des végétaux désertiques et subdésertiques de l'Algérie. Bull Soc Bot Fr 80:775
Killian CH, Faurel L (1936) La pression osmotique des végétaux du Sud Algérien. Ann Physiol 12:5
Killian CH, Lemee G (1956) Les zerophytes: leur economie d'eau. In: Ruhland W (ed) Handbuch der Pflanzenhyiologie, vol 3. Springer, Berlin Heidelberg New York, pp 787–824
Kluge M (1976) Carbon and nitrogen metabolism under water stress. In: Lange OL, Kappen L, Schulze E-D (eds) Ecological studies, vol 19. Springer, Berlin Heidelberg New York, pp 243–252
Kneucker A (1926) Eine neue Flechte von Sinai und vom Ufer des Toten Meeres. Allg Bot Z 30/31:43
Korstian GF (1924) Density of cell sap in relation to environmental conditions in the Wasatch mountains, Utah. J Agr Res 28:845–907
Kozlowski J, Wieger R G (1986) Optimal allocation of energy to growth and reproduction. Theor Popul Biol 29:16–37
Kozlowski TT (1964) Water metabolism in plants. Harper & Row, New York
Kramer PJ (1946) Absorption of water through suberized roots of trees. Plant Physiol 21:37–41
Kramer PJ (1980) Plant and soil relationships: a modern synthesis. McGraw-Hill, New York
Kreeb KH (1963) Hydrature and plant production. In: Rutter AJ, Whitehead EH (eds) The water relations of plants. Blackwell, Oxford, pp 272–288
Kylin A, Gee R (1970) Adenosine triphosphatase activities in leaves of the mangrove *Avicennia nitida* Jacq. Plant Physiol 45:169
Laetsch WLM (1974) The C4 syndrome: a structural analysis. Annu Rev Plant Physiol 25:27–52

Lange OL (1965a) Leaf temperatures and method of measurement. Proceedings of the Montpellier Symposium, Methodology of plant eco-physiology, UNESCO, pp 203–209

Lange OL (1965b) Der CO_2 Gaswechsel von Flechten nach Erwärmung im feuchten Zustand. Ber Dtsch Bot Ges 78:441–454

Lange OL (1969a) Die funktionellen Anpassungen der Flechten an die oekologischen Bedingungen arider Gebiete. Ber Dtsch Bot Ges 82:3–22

Lange OL (1969b) Ecophysiological investigations in lichens of the Negev desert. I. CO_2 gas exchange of *Ramalina macriformis* (Del.) Bory under controlled conditions in the laboratory. Flora 158:324–359

Lange OL, Evenari M (1971) Experimentell-ökologische Untersuchungen an Flechten der Negev-Wüste. IV. Wachstumsmessungen an *Caloplaca aurantia* (Pers) Hellb. Flora 160:100–104

Lange Ol, Ziegler H (1986) Different limiting processes of photosynthesis in lichens. In: Marcelle R, Clijsters H, Van Poucke M (eds) Biological control of phtosynthesis. Martinus Nijhoff, Dordrecht, pp 147–161

Lange OL, Koch W, Schulze ED (1969) CO_2-Gaswechsel und Wasserhaushalt von Pflanzen in der Negev-Wüste am Ende der Trockenheit. Ber Dtsch Bot Ges 82:39–61

Lange OL, Schultze E-D, Koch W (1970a) Experimentell-ökologische Untersuchungen an Flechten der Negev-Wüste. II. CO_2 – Gaswechsel und Wasserhaushalt von *Ramalina maciformis* (Del.) Bory am natürlichen Standort während der sommerlichen Trockenperiode. Flora (Jena) 159:38–62

Lange OL, Schultze E-D, Koch W (1970b) Experimentell-ökologische Untersuchungen an Flechten der Negev-Wüste. III. CO_2 – Gaswechsel und Wasserhaushalt von Krusten-und Blattflechten am natürlichen Standort während der sommerlichen Trockenperiode. Flora (Jena) 159:525–538

Lange OL, Lösch R, Schulze E-D, Kappen L (1971) Responses of stomata to change in humidity. Planta 100:76–86

Lange OL, Schulze E-D, Kappen L, Buschbom U, Evenari M (1975) Adaptations of desert lichens to drought and extreme temperatures. In: Hadley NF (ed) Environmental physiology of desert organisms. Dowden, Hutchinson and Ross, Stroudsberg, pp 20–37

Le Houérou HN (1982) An outlne of the bioclimatology of Libya. Bull Soc bot Fr 131. Actual Bot 2/4/4:157–178

Lemée G (1953) Copntributions a l'étude écologique de la végétation des confins Saharo-marocains. Desert Research. Proc Int Symp Jerusalem, pp 302–306

Lemée G (1954) L'économie de l'eau chez quelque Graminée vivaces du sahara septentrional. Vegetatio 5/6:539–541

Levitt J (1956) The hardiness of plants. Academic Press, London

Levitt J (1972) Responses of plants to environmental stresses. Academic Press, London

Levitt J (1976) Physiologoical basis of stomatal response. In: Lange OL, Kappen, Schulze E-D (eds) Water and plant life: problems and modern approaches. Springer, Berlin Heidelberg New York, pp 160–168

Levitt J (1980) Responses of plants to environmental stresses, vol I: chilling, freezing, and high temperatures. Academic Press, New York

Levitt L, Sullivan CY, Krull E (1960) Some problems in drought resistance. Bull Res Counc Isr Biol 12:207–219

Liphschitz N, Waisel Y (1974) Existence of salt glands in various genera of the *Gramineae*. New Phytol 73:507–513

Liphschitz N, Waisel Y (1982) Adaptation of plants to saline environments: salt excretion and glandular structure. In: Sen DN, Rajpurhit KS (eds) Tasks for vegetation science. Junk, The Hague

Lloyd FE (1905) The artificial induction of leaf formation in the ocotillq. Torreya 5:175–179

Lösch R (1989) Plant water relations. Prog Bot 50:27–50

Lüttge U (1971) Structure and function of salt glands. Annu Rev Plant Physiol 22:23–44
Mahmoud MI (1965) Protoplasmics and drought resistance in mosses. PhD Thesis, Univ California, Davis
Marloth R (1887) Zur Bedeutung der salzabscheidenden Drüsen der Tamariscineen. Ber Dtsch Bot Ges 5:319–324
Marsh FL (1940) Water content and osmotic pressure of certain prairie plants in relation to environment. Univ Nebr Stud XL(3)
Maun MA (1996) The effects of burial by sand on survival and growth of *Calamovilfa longifolia.* Ecoscience 3:93–100
Maun MA, Lapierre J (1984) The effects of burial by sand on *Ammophila breviligulata.* J Ecol 72:827–839
Maximov NA (1931) The physiologoical significance of xeromorphic structure of plants. J Ecol 19:273–282
Mendelssohn IA (1987) Effect of salinity on proline accumulation in three *Sparlina* species. Proc 14th Int Bot Congr, Abstr no 1-22a-6
Migahid AM (1938) Binding of water in relation to drought resistance. Bull Fac Sci Found I Univ 18:5–28
Migahid AM (1962) The drought resistance of Egyptian desert plants. Proc Arid Zone Symp Plant-Water Relations, UNESCO, Madrid, pp 213–233
Migahid AM, Abdel Rahman AA (1953) Observations on the drought resistance of desert plants. Bull Inst Desert d'Égypte 3:58–83
Migahid AM, Abdel Wahab AM, Batanouny KH (1972) Eco-physiological studies on desert plants. VII. Water relations of *Leptadenia pyrotechnica* (Forssk.) Decne. growing in the Egyptian desert. Oecologia (Berl) 10:79–91
Migahid AM (1988–1989) Flora of Saudi Arabia, 3rd edn. King Saud University Libraries, Riyadh, Saudi Arabia
Monod T (1954) Modes contractés et diffusè de la vegetation Saharienne. In: Cloudsley-Thompson JL (ed) Biology of desert. Institute of Biology, London, pp 35–44
Montasir AH, Migahid AM (1934) Transpiration and stomata in desert plants. Bull Fac Sci Found I Univ (Cairo) 1:1–33
Moss EH (1940) Interxylary cork in Artemisia with a reference to its taxonomic significance. Am J Bot 27:762–768
Mozafer A, Goodin JR (1970) Vesiculated hairs: a mechanism for salt tolerance in *Atriplex halimus* L. Plant Physiol 45:62–65
Müller J (1884) *Enumerationis lichenum aegypiacoreum.* Suppl I Rev Mycol 21:3–9
Müller J (1891) Lichenologische Beiträge. Flora 74:371–382
Nasr Seyyed Hussein (1976) Islamic science; an illustrated study. World of Islam Festival Publ, London
Negbi M, Evenari M (1962) The means of survival of some desert summer annuals. Proceedings of the Madrid Symposium. Arid zone research, vol XVI. Plant-water relationship in arid and semi-arid conditions. UNESCO, Paris pp 249–259
Newman EI (1976) Water relations. In: Hall M A (ed) Plant structure, of *Calamovilfa longifolia.* Ecoscience 3:93–100
Newman EI (1978) Water relations. In: Hall MA (ed) Plant structure, function and adaptation. Macmillan Press, London, pp 157–196
Nobel Park S (1996) Ecophysiology of roots of desert plants, with special emphasis on agaves and cacti. In: Waisel Y, Eshel A, Kafkafi U (eds) Plant roots: the hidden half, 2nd edn. Marcel Dekker, New York, pp 823–844
Noy-Meir I (1986) Desert ecosystems: structure and function. In: Evenari M, Noy-meir I, Goodall DW (eds) Hot deserts and arid shrublands. Elsevier, Amsterdam, pp 93–103
Nylander W (1864) Lichens in Aegypto a cel Ehrenberg collecti. Act Soc Linn de Bordeau 25:59–66

Onal M (1965) Beiträge zum Halophyten-Problem. Ber Dtsch Bot Ges 78 (2):68–72
Onal M (1966) Vergleichende ökologische Untersuchungen bei Halophyten und Glycophyten in der Nähe von Neapel. Rev Fac Sci Univ Istanbul Ser B 31:209–248
Oppenheimer HR (1932) Zur Kenntnis der hochsommerlichen Wasserbilanz Mediteraner Gehölze. Ber Dtsch Bot Ges 50a:185–245
Oppenheimer HR (1947) Studies on the water balance of unirrigated woody plants. Pal J Bot Rehovot Ser 6:63–77
Oppenheimer HR (1951) Summer drought and water balance of plants growing in the Middle East. J Ecol 39:356–362
Oppenheimer HR (1960) Adaptation to drought: xerophytism. Arid zone research, no 15. Plant-water relationships in arid and semi-arid conditions. Review of research, UNESCO, Paris, pp 105–138
Oppenheimer HR, Halevy AH (1962) Anabiosis of *Ceterach officinarum* Lam. ex DC. Bull Res Counc Isr D3 11:127–147
Orshan G (1953) Note on the application of Raunkiaer's system of life forms in arid regions. Palestine J Bot Jerusalem Ser 6:120–122
Orshan G (1963) Seasonal dimorphism of desert and Mediterranean chamaephytes and its significance as a factor in their water economy. In: Rutter AJ, Whitehead FH (eds) The water relations of plants. Wiley, New York, pp 206–222
Orshan G (1986) Plant form as describing vegetation and expressing adaptation to environment. Ann Bot (Roma) 44:7–38
Orshan G, Zand G (1962) Seasonal body reduction of certain desert halfshrubs. Bull Res Counc Isr 11D(1):35–42
Osmond CB, Bjorkman 0, Anderson DJ (1980) Physiological processes in plant ecology toward a synthesis with *Atriplex*. Ecological studies, vol 36. Springer, Berlin Heidelberg New York
Osmond CB, Winter K, Ziegler H (1982) Functional significance of different pathways of CO_2 fixation in photosynthesis. In: Lange OL, Nobel PS, Osmond CB, Ziegler H (eds) Encyclopedia of plant physiology, new series, vol 12B. Physiological plant ecology II. Springer, Berlin Heidelberg New York
Osonubi O, Davies WJ (1981) Root growth and water relations of oak and birch seedlings. Oecologia (Berl) 51:343–350
Parker J (1968) Drought-resistance mechanisms. In: Kozlowski TT (ed) Water stress in plants. Academic Press, New York, pp 195–234
Passioura JB (1982) Water in the soil-plant – atmosphere continuum. In: Lange OL, Nobel PS, Osmond CB, Ziegler H (eds) Encyclopedia of plant physiology. New series, vol 12B. Physiological plant ecology. Springer, Berlin Heidelberg New York, pp 5–33
Pollak G, Waisel Y (1970) Salt secretion in *Aleuropus littoralis* (Willd.) Parl. Ann Bot (Lond) 34(137):879–888
Pollak G, Waisel Y (1973) Ecophysiology of salt excretion in *Aleuropus littoralis* (Gramineae). Physiol Plant 47:177–184
Price JR (1938) The roots of some north African desert grasses. New Phytol 10:328–340
Prusakova LD (1960) Influenceof water relations on tryptophan synthesis and leaf growth in wheat. Fiziol Rast 7:139–148
Raschke K (1956) Über die physikalischen Beziehungen zwischen Wärmeübergangszahl, Strahlungsaustausch, Temperatur und Transpiration eines Blattes. Planta 48:200–238
Rains DW, Epstein E (1967) Preferential absorption of potassium by leaf tissue of the mangrove *Aticennia marina*: an aspect of halophytic competence in coping with salt. Aust J Biol Sci 20:847–857
Raunkiaer C (1934) Life forms of plants and statistical plant geography (collected papers translated into English). Oxford University Press, Oxford

Reader RJ, Jalili A, Grime JP, Spencer RE, Matthews N (1992) A comparative study of plasticity in seedling rooting depth in drying soil. J Ecol 81:543–550
Reichenberger G, Pyke DA (1990) Impact of early root competition on fitness components of four semi-arid species. Oecologia (Berl) 85:159–166
Reeve HK, Sherman PW (1993) Adaptation and the goals of evolutionary research. Q Rev Biol 68:1–32
Repp G (1961) The salt tolerance of plants; basic research and tests. UNESCO Arid Zone Res 14:153–161
Reynolds JF (1986) Adaptive strategies of desert shrubs with special reference to the creosotebush (*Larrea tridentata* (DC) Cov.). In: Whitford WG (ed) Pattern and process in desert ecosystem. Univ of New Mexico Press, New Mexico, pp 19–49
Rhoades MM, Carvalho A (1944) The function and structure of the parenchyma sheath plastids of the maize leaf. Bull Torr Bot Club 71:335–346
Ried A (1960) Thallusbau und Assimilationshaushalt von Laub- und Krustenflechten. Biol Zentralbl 79:129–151
Rogers RW (1971) Distribution of the lichen *Chondropsis semiviridis* in relation to its heat and drought resistance. New Phytol 7:1069–1077
Roschall E (1938) Eine physiologische Studie an *Ceterach officinarum* Willd. Flora 132:305–318
Routley DG (1966) Proline accumulation in wilted ladino clover leaves. Crop Sci 6:358–361
Rozema J (1975) An eco-physiological investigation into the salt tolerance of *Glaux maritima* L. Acta Bot Neerl 24:407–417
Ruhland W (1915) Untersuchungen über die Hautdrüsen der Plumbaginacean. Ein Beitrag zu den Halophyten. Jahrb Wiss Bot 55:409–498
Rundel PW (1978) Ecological relationships of desert fog zone lichens. Bryologist 81:277–293
Rundel PW (1982) Water uptake by organs other than roots. In: Rundel PW, Nobel PS (1991) Structure and function in desert root system. In: Atkinson D (ed) Plant root growth-an ecological prospective. Blackwell, Oxford, pp 349–378
Schobert B (1979) Die Akkumulierung von Prolin in *Phaedactylum tricornutum* und die Funktion der "compatible solutes" in Pflanzenzellen unter Wasserstress. Ber Dtsch Bot Ges 92:23–30
Schobert B (1980) The function of proline accumulation in the diatom *Phacodactylum tricornutum* during water stress. In: Spanswick RM, Lucas WJ, Danity J (eds) Plant membrane transport. Elsevier/North-Holland Biomedical Press, Amsterdam, pp 487–488
Scholander PF, Hammel HT, Hemmingsen EA, Garey W (1962) Salt balance in mangroves. Plant Physiol 37:722–729
Schrader HA (1809) Über Palla's Halophyta mit besonderer Rücksicht auf die Gattungen *Salsola* und *Suaeda*. Schrad Neues J Bot 3:58–92
Schtscherback J (1910) Über die Salzausscheidung durch die Blätter von *Statice gtnelitzi.* Ber Dtsch Bot Ges 28:30–34
Schulze E-D (1982) Plant life forms and their carbon, water and nutrient semi-arid regions. Annu Rev Ecol Syst 17:89
Schultze E-D, Hall A, Lange OL, Evenari M,Kappen L, Buschbom U (1980) Long term effects of drought on wild and cultivated plants in the Negev Desert. 1. Maximal rates of net photosynthesis. Oecologia (Berl) 45:11–18
Seybold A (1930) Untersuchungen über die Transpirations-Widerstände und über die Temperaturen ägyptisch-arabischer Wüstenpflanzen. Planta 9:270–314
Shields LM (1950) Leaf xeromorphy as related to phyiological and structural influences. Bot Rev 16:299–447
Singh TN Aspinall D, Paleg LG, Boggess SF (1973a) Stress metabolism. 11. Changes in proline concentration in excised plant tissue. Aust J Biol Sci 26: 57–63

Singh TN, Paleg LG, Aspinall D (1973b) Stress metabolism. 111.Variations in response to water deficit in the barley plant. Aust J Biol Sci 26:64–76
Slatyer RD (1956) Absorption of water from atmospheres of different humidity and its transport through plants. Aust J Bot 9:552–558
Slatyer RD (1960) Aspects of the tissue water relationships of an important arid zone species (*Acacia aneura* F. Mull) in comparison with two mesophytes. Bull Res Counc Isr Sect D8: 159–168
Slatyer RD (1967) Plant-water relationships. Academic Press, London
Smith DC (1962) The biology of lichen thalli. Biol Rev 37:537–570
Smith DC (1975) Symbiosis and the biology of lichenised fungi. In: Symbiosis. Symp Soc Exp Biol 72:373–405
Sober E (1984) The nature of selection. MIT Press, Cambridge
SpauldingVM (1905a) Biological relationships of desert shrubs. II. Absorption of water by leaves. Bot Gaz (Chicago) 41:262–282
SpauldingVM (1905b) Absorption of atmospheric moisture by desert shrubs. Bull Torr Bot Club 33:367–375
Stark N, Love LD (1969) Water relations of three warm desert species. Isr J Bot 18:175–190
Stewart CR. Morris CJ, Thompson JF (1966) Changes in amino acid content of excised leaves in the dark, Plant Physiol 41:1585–1590
Stewart CR (1972) Proline content and metabolism during rehydration of wilted excised leaves in the dark. Plant Physiol 50:679–681
Stewart CR (1973) The effect of wilting on proline metabolism in excised bean leaves in the dark. Plant Physiol 51:679–681
Stewart CR (1981) Proline accumulation: biochemical aspects. In: Paleg LG, Aspinall D (eds) The physiology and biochemistry of drought resistance in plants. Academic Press, Sidney
Stewart CR, Lee JA (1974) The role of proline accumulation in halophytes. Planta 120:279–289
Stocker O (1928a) Der Wasserhaushalt ägyptischer Wüsten- und Salzpflanzen. Fischer, Jena, 200 pp
Stocker O (1928b) Das Halophyten-Problem. Ergeb Biol 3:265–353
Stocker O (1929) Das Wasserdefizit von Gefaesspflanzen in verschiedene Klimazonen. Planta 7:282–287
Stocker O (1933) Salzpflanzen. Handb Naturwiss 8:699–712
Stocker O (1956) Die Dürreresistenz. In: Rhuland W (ed) Encyclopedia of plant physiology, vol 3. Springer, Berlin, Heidelberg, New York, pp 696–741
Stocker O (1960) Physiological and morphological changes in plants due toi water deficiency. In: Plant-water conditions, vol 15. UNESCO, Paris, pp 63–104
Strogonov BP (1964) The physiological basis of salt tolerance of plants (translated from Russian). Isr Res Counc, Jerusalem
Sveshnikova VM (1972) On the absorption of water vapour by the overground parts of the Kara-Kum Desert plants (in Russian). Bot Zh (Leningrad) 57:880–886
Thompson JF, Stewart CR, Morris CJ (1966) Changes in amino acid conent of excised leaves during incubation I. The effect of water content of leaves and atmospheric oxygen level. Plant Physiol 41:1578–1584
Tilman D (1988) Plant strategies and the dynamics and structure of plant communities. Princeton University Press, Princeton
Tyree MT, Jarvis PG (1982) Water in tissue and cells. In: Lange OL, Nobel PS, Osmond CB, Ziegler H (eds) Encyclopedia of plant physiology. New series, vol 12B: physiological plant ecology. Springer, Berlin Heidelberg New York, pp 35–77
Ungar IA (1965) An ecological study of the vegetation of the Big Salt Marsh, Stafford Country, Kansas. Univ Kan Sci Bull 46(1):1–99

Ungar IA (1967) Influence of salinity and temperature on seed germination. Ohio J Sci 67: 120–123

Van Eijk M (1939) Analyse der Wirkung des NaCl auf die Entwicklung, Sukkulenz und Transpiration bei *Salicornia herbacea,* sowie Untersuchungen über den Einfluss der Salzaufnahme auf die Wurzelatmung bei *Aster triloba* Rec. Trav Bot Neerl 36:559–657

Vesey-Fitzgerald DF (1957) The vegetation of central and eastern Arabia. J Ecol 45:779–793

Viera da Silva (1976) Water stress, ultrastructure and enzymatic activity. In: Lange OL, Kappen L, Schulze (eds) Water and plant life. Problems and modern approaches. Springer, Berlin Heidelberg New York, pp 207–224

Volkens G (1884) Die Kalkdrüsen der Plumbaginacean. Ber Dtsch Bot Ges 2:334–342

Volkens G (1887) Die Flora der ägyptisch-arabischen Wüste auf Grundlage anatomisch-physiologischer Forschungen dargestellt. Gebrüder Borntraeger, Berlin, 156 pp+18 plates

Waisel Y (1958) Dew absorption by plantsof arid zones. Bull Res Counc Isr Ser D 6:180–186

Waisel Y (1961) Ecological studies on *Tamarix aphylla* (L.) Karst. III. The salt economy. Plant Soil 13:356–364

Waisel Y (1963) Ecotypic differentiation in the Flora of Israel. III. Anatomical studies of some ecotype pairs. Bull Res Counc Isr 11D:183–190

Waisel Y (1972) Biology of halophytes. Academic Press, New York

Waisel Y (1989) Screening for salt resistance. Proc 21st Coll Int Potash Inst, Bem, pp 143–155

Waisel Y, Pollak G (1969) Estimation of water stresses in the active root zone of some native halophytes in Israel. J Ecol 57:789–794

Waisel Y, Gude H, Pollak G (1981) An ecophysiological study of the salt secretion of four halophytes. New Phytol 89: 201–217

Waisel Y, Eshel A, Agami M (1986) Salt balance of leaves of the mangrove, *Avicennia maritia.* Physiol Plant 67:67–72

Walter H (1937) Die Ökologischen Verhältnisse in der Namib Nebenwüste (Südwestafrika). Jahrb Wiss Bot 84:58–219

Walter H (1955a) Die Klimadiagramme als Mittel zur Beurteilung der Klimaverhältnisse für ökologische, vegetationskundliche und landwirtschaftliche Zwecke. Ber Dtsch Bot Ges 68: 331

Walter H (1955b) The water economy and the hydration of plants. Annu Rev Plant Physiol 6:239–252

Walter H (1957) Klimadiagramme als Grundlage zur Feststellung von Dürrezeiten. Wasser Nahrung 1:8–11

Walter, H (1963) The water supply of desert plants. In: Rutter AJ, Whitehead EH (eds) The water relationships of plants. British Ecological Society, London, pp 199–205

Walter H (1964) Die Vegetation der Erde in ökophysiologischer Betrachtung, vol I. Gustav Fischer, Jena

Walter H (1973) Vegetation of the earth in relation to climate and the ecophysological conditions. Engl Univ Press, London/Springer, Berlin Heidelberg New York

Walter H (1974) A new approach to the water relations of desert plants. In: Brown GW Jr (ed) Desert biology, vol II. Academic Press, New York, pp 213–310

Walter H, Lieth H (1959) Klimadiagramme Weltatlas. Gustav Fischer, Jena

Walter H, Steiner M (1936) Die Ökologie der ostafrikanischen Mangroven. Z Bot 30:65–193

Weissenbock G (1969) Einfluss des Bodensalzgehaltes auf Morphologie und Ionenspeicherung von Halophyten. Flora 158:369–389

Went FW (1942) The dependence of certain annual plants on shrubs in southern California deserts. Bull Torr Bot Club 69:100–114

Williams MC (1960) The effect of sodium and potassium salts on growth and oxalate content of *Halogetoti.* Plant Physiol 35(4):500–505

Wilson MF (1983) Plant reproductive ecology. Wiley, New York

Wilson JR, Ludlow MM, Fisher MJ, Schultze E-D (1980) Adaptation to water stress of the leaf water relations of four tropical forage species. Aust J Physiol 7:207–220
Winter K (1979) Photosynthetic and water relationships of higher plants in saline environment. In: Jefferies RL, Davy AJ (eds) Ecological processes in coastal environments. Blackwell, Oxford, pp 297–320
Winter K, Ziegler H (1982) Functional significance of different pathways of CO_2, fixation in photosynthesis. In: Lange OL, Nobel PS, Osmond CB, Ziegler H (eds) Encyclopedia of plant physiology. new series, vol 12B. Physiological plant ecology II. Springer, Berlin Heidelberg New York
Wood JG (1923) On transpiration in the field of some plants from the and portions of South Australia, with notes on their physiological anatomy. Trans R Soc S Aust (Adelaide) 47: 259–278
World Resources Institute (1996) World resources: a guide to the global environment. United Nations Environmental Program (UNEP), United Nations Development Program (UNDP), The World Bank. Oxford University Press, New York
Zaghloul M (1928) The soil stratometer. A method for the examination of deep lying soil. Nature (April 7):337
Zayed M (1984) Ecological studies on *Retama raetam* in the Egyptian desert. MSc Thesis, Fac Science, Cairo University
Zhang J, Maun MA (1992) Effects of burial in sand on the growth and reproduction of *Cakile edentula*. Ecography 15:296–302
Ziegler H, Lüttge U (1966) Die Salzdrüsen von *Limonium vulgare*. 1. Die Feinstruktur. Plantas 70:193–206
Ziegler H, Batanouny KH, Sankhla N, Vyas O, Stichler W (1981) The photosynthetic pathway of some desert plants from India, Saudi Arabia, Egypt, and Iraq. Oecologia (Berl) 48:93–99
Zohary M (1962) Plant life in Palestine, Israel and Jordan. Ronald Press, New York, 262 pp
Zohary M (1973) Geobotanical foundations of the Middle East. Gustav Fischer, Stuttgart
Zohary M, Orshan G (1954) Ecological studies in the vegetation of the Near East Desert. V. The *Zygophylletum dumosi* and its hydro-ecology in the Negev of Israel. Vegetatio 5/6:341–350

Index of Latin Plant Names

Index of Arabic Plant Names

Subject Index

MIX
Papier aus verantwortungsvollen Quellen
Paper from responsible sources
FSC® C105338
FSC
www.fsc.org

If you have any concerns about our products,
you can contact us on
ProductSafety@springernature.com

In case Publisher is established outside the EU,
the EU authorized representative is:
Springer Nature Customer Service Center GmbH
Europaplatz 3, 69115 Heidelberg, Germany

Printed by Libri Plureos GmbH
in Hamburg, Germany